THE SCIENCE OF LIFE:
THE LIVING SYSTEM—
A SYSTEM FOR LIVING

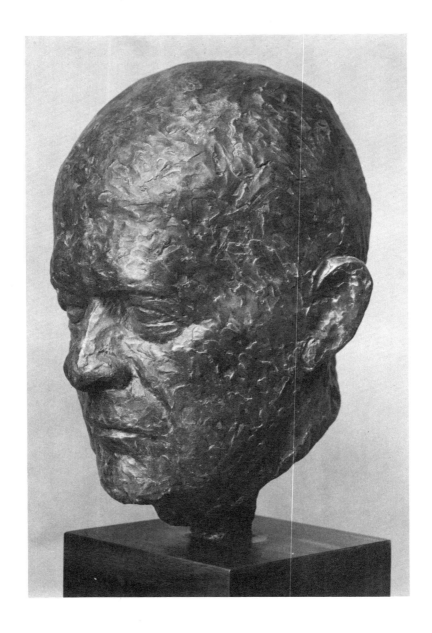

Frontispiece: From a sculpture by Vera Kuffner Eberstadt.

THE SCIENCE OF LIFE:
THE LIVING SYSTEM—
A SYSTEM FOR LIVING

PAUL A. WEISS
THE ROCKEFELLER UNIVERSITY, NEW YORK, N.Y.

FUTURA
PUBLISHING
COMPANY
1973

Copyright © 1973,
by Futura Publishing Company, Inc.

Published by
Futura Publishing Company, Inc.
295 Main Street
Mount Kisco, New York 10549

L.C. #: 73-85507

ISBN #: 0-87993-034-9

All rights reserved.
No part of this book may be translated or reproduced in any form without the written permission of the publisher.

Printed in U.S.A. by
NOBLE OFFSET PRINTERS, INC.
New York, N.Y. 10003

ACKNOWLEDGEMENTS

The extensive experimental work of the author in cellular, developmental, and neural biology (from 1918 to the present), from which the generalizations and lessons of this book have been distilled, has been most generously supported off and on, besides from the Institutions and Universities in which he served, by the following: Academy of Sciences, Vienna; Notgemeinschaft der Deutschen Wissenschaften; Abbott Memorial Fund, University of Chicago; American Philosophical Society; American Academy of Arts and Sciences, Boston; Rockefeller Foundation; U.S.A. Office of Scientific Research and Development; National Foundation for Infantile Paralysis; American Cancer Society; Faith Foundation, Houston; National Institutes of Health (U.S. Public Health Service); National Science Foundation. To them all he expresses his deep gratitude for having enabled him to combine work and cogitation in a steady course of mutual reinforcement.

CONTENTS

FOREWORD

THE ORDER OF LIFE

Prologue: And What Is Life?	1
Life Science: Public Servant	5
A Verbal Trap: "Genetic Determinism"	7
The Organism: A Set of Chinese Boxes	10
The Egg: A Patterned Organism	13
Determinism Stratified	18
Macrodeterminacy	20
A Canon for Determinacy	22
Determination: Meaning What?	24
Hierarchic Systems: Measured by "Grain Size"?	26
Field Patterns: Properties of Collectives	29
Systems: Theoretically Founded	35
The Living Cell: A System	38
Systems: Operationally Defined	40
Organic Form: The Field-Gene Dualism	42
Cell Populations: Models of Freedom	49
Free Systems: Problem Solvers	52

Consolidation: Systems Becoming Mechanisms 55
Survival: Balance Between Rigor and Plasticity 57
What Genes Determine: Substantives or Attributes? ... 61
Life: Matter or Process? 65
The "Origin of Life": Facts and Fancies 68
Primordial Life: Test of Primordialism 75
Human Fate: From Mechanism to Fatalism? 78
Breaking the the Sound Barrier: Beyond Genetics—Epigenetics 80
Fitness from Stress: Man's Adaptive Scope 83
"Freedom of Choice": The Brain as System 85
Man's Mission: Universality Amidst Diversity 89
Life Science's Lesson: The Primacy of Order 91
Epilogue: So, Finally, What is Life? 94

POSTSCRIPT: FROM LIFE SCIENCE TO EDUCATION FOR LIVING

Education: Guide to Harmony 96
"Linear Causality": Predicament of Man? 99
Category Versus Continuum: Conceptual Confrontation? 106
Towards "Depolarization": A Task for Education 109
Examining the Educational Process 111
(1) What Is the Purpose of Education? 111
(2) Self-Development and Universal Education: Another Confrontation? 113
(3) Furthering Self-Development 115
(4) Liberal Education and Vocational Training: Further Confrontations? 117

Individuality within the Limits of "Context" 121
Group and Individual: Does "Average Man" Exist? 124
Time Course of Life: Eternal Change 127

EPILOGUE 131
REFERENCES 136

FOREWORD

A wave may depict fluctuations of our physical environment or of an economic cycle or of the moods and motives of the minds of men, swinging upward and downward on and on, about a median band of stationary equilibrium. When left alone, waves will subside, as in the calming of the sea after the storm, the damping of an oscillating string, the cooling off of tempers after quarreling, the trend to compromise in bargaining. This is the natural course of natural events in *inorganic* nature and of conciliatory human behavior. Thermodynamics teaches us that without reinforcement the swing loses motive force inexorably, as energy to entertain it is dissipated as tribute to the growing pool of entropy. But there are instances in which that downdrift is temporarily reversed. They are a prime distinction of *living* systems. An organism can borrow dissipation-bound energy in one place or period to spend it at another place or time to feed movement or other energy-requiring work. A living being thus not only can keep a wave in motion, but properly timed, can make it swell. Yet, to profit from this game with nature, man must know its rules and play it right. For instance, stored energy, when added to the upstrokes only of a rhythmic motion, makes the momentum of the whole swing grow. Man's productive and creative faculties, but also his powers of destruction, hinge on the ways in which he exploits this principle; for as it enables even frail men to toll a heavy bell, it also makes it possible for political villains to foment vicious mob action by the reiterative reinforcement of an insidious rumor. In fact, even erstwhile beneficial excursions can become ruinous when driven to exceed a given tolerable range; for then the following is bound to happen.

Any patterned living system—be it a cell, an organism, a

community, or a society—is held together as a viable integrated entity by inner ties that are elastic; the system yields to moderate distortion, and the greater the distortion, the greater is also the counterforce that contains it within bounds. Elastic strain, however, can be sustained only up to the stress limit. If stretched beyond, the bonds will snap: the system loses its cohesion, crumbles, flies apart. Disjointed, the separated pieces then carry on, like moons, as isolated entities in mutual opposition, no longer even tenuously connected. Gone are the connecting threads that kept the parts united as a system. Just think of whirling an object on a rubber string round and round at increasing speed until centrifugal force breaks the string and the object flies off. This is the way in which antagonisms give rise to schisms: by promoting centrifugal disruption. Our world is full of examples. Issues are artifically polarized, the spotlight being turned on the extremes, and then the tug-of-war between them goes on for good. The playful spirit of win-or-lose of the sports field, with friendly handshakes at the end, is violently perverted into a do-or-die precept for either-or survival. What have been merely extreme ends of a continuous spectrum of positions along a graded scale of values become implacably hostile camps contesting for monopolistic status: antagonisms that have started from sheer accentuation of extreme points of view deteriorate into factual conflict and combat, and people change from occupants of opposite ideological stations into actual aggressive opponents.

This tendency for antipolar extremes to rise through emphasis and prominence to eventual total separation is deeply ingrained in the nature of biological existence. Biological nature condones the resolution of conflict by victory rather than by conciliation. Outfighting, outbreeding, or outsmarting a competitor are approved methods of evolutionary progress; true, there are also instances in nature of cooperation and harmonization of interests, as for instance in symbiosis, but those are essentially accommodative arrangements among noncompetitive groups. Being the animal he is, man has inherited a flair for polarizing issues. Instinctively, he even fans their conflict-breeding potential by laying stress on divergence and disparity; doing so, he amplifies the centrifugal separative forces which threaten to disrupt the crucial cohesiveness without which no living system, including the human race, can survive. In short, *biological* man cannot be trusted to act in the best interest of humanity.

But then, man is much more than an animal. Through his power of reasoning, he can, if not abolish his deep-set instincts, at least repress and supersede them by intelligent action, whenever he recognizes his primitive instinctive habits to run counter to the higher ideals of a civilized society. Rational foresight lets him spot danger signals of strain approaching stress limit, and insight, gained from experience, enables him to take countermeasures in time to forestall imminent disaster by brakelike damping, decelerating and draining the motive force of extreme swings. In principle, such deliberate regulatory intervention by man corresponds, of course, to the automatic "negative feedback" devices to which any self-preserving system, living or nonliving, owes its capacity for maintaining, or oscillating about, a state of relative stability of pattern. Yet, human corrective counteractions differ from those of automatons in their lack of built-in self-restraints: a human counterthrust started as a control act often ends up losing its own self-control.

That is to say, man, although he clearly recognizes the direction in which to counter an excessive move, is not equally adept at choosing the right amplitude for the checking force. A faulty sense of perspective often makes him underrate or overrate it. Misjudgement and timidity lead him to undercorrect, while overreaction to real or envisaged outrage prompts him to overcompensate. There are those who close their eyes to controversy by self-delusion, by simply disputing that the professed extremes actually exist; and there are those others who, in their violent aversion to an excessive swing in one direction, reverse the wave into one no less extreme, even though of opposite sign.

Now, let me try to explain what that metaphoric—and somewhat cryptic—preamble on wave dynamics has to do with this critical study of life. I used the simile of undulatory patterns to symbolize man's groping for a civilized existence; more specifically, to emphasize two basic, though commonplace, observations: (1) that *biological* man tends to create antagonisms by emphasizing the extreme ends of a continuous spectrum of notions composed of truths, as well as fallacies and prejudices; (2) that *civilized* man, although he has the rational endowment to counteract that fatal trend to excessive polarization, seems not to have as yet learned to make the most of his rational faculties by adopting the broad perspective which would let moderation defuse the explosive charge of fanatical antagonisms.

Perhaps more than perspective is needed. What seems to have been lacking above all has been a judicious mixture of idealism and realism, that is, of theory and practice. In theory, one can set for each truthful proposition on opposite, which obviously, by definition, is untrue. Nature offers us models of such absolute bipolarity in positive and negative electricity and the antipoles of magnetism. But the real world of human affairs, in which we live, which we observe, and in which we then look for principles that would merit the accolade of "pure truth", has yet to bring forth a single example of such ideal unadulterated truth. Consequently, we must not expect to meet in the real human world the antithetical phenomenon of "unmitigated untruth", either. The idealistic-realistic precept of civilized society, therefore, should be to sort from both extremes the positive "truthful" fractions which they contain and rally them into a cohesive system, the unity of which would avert irreparable schisms. Thus, what the popular precept of "accentuating the positive" espouses, might lead, in a more sophisticated version, to the depolarization of exaggerated antipolarities, and thereby to the conciliation of putatively irreconcilable antagonisms: in short, to greater harmony through the moderating influence of the broader views gained from a balanced perspective.

To help ascend to such broad perspective should become a prime task of all systems of education and public enlightenment. Inspected soberly, any two phenomena or ideas one might wish to compare critically present both common, or "generic", and distinctive, or "specific", features. As the generic ones, in their repetitiveness, become dull, the specific ones, more conspicuous for being more unique, attract attention and monopolize our interest, thus putting in eclipse the stable steady bulk of common features that unite them.

Minor disparities outshine major identity. The picture of the world one is apt to get from such a confinement of the field of vision will of necessity be a grossly lopsided one. As viewpoints have a way of consolidating into standpoints, lopsided views become the bases for lopsided decisions and actions. Moreover, the habit of singling out for prime attention and categorical confrontation the opposite ends of a continuous scale of values is just a special case of a more general fallacy; namely, that of concentrating on the differences among a set of subjects while ignoring the much larger substance they have in common, of which the differentials are but abstracted attributes.

This is a state of affairs in which education should assume a major corrective and preventive function, hopefully to be joined by the media of public communication. Their task would be to break the obsessive habit of focusing compulsively on single isolated issues, separated from their context—a major source of partisan self-confinement—and develop in its stead the habit of letting the mind sweep back and forth over the whole continuum of phenomena, which constitute the context. This evidently would imply a major change of attitude, a reorientation of viewpoint and focus of thoughts from centrifugal decomposition of something that has unity to centripetal convergence.

Such a changeover from primitive accent on opposites to civilized practice of moderation looks like a gigantic job; and so it is. It reminds one of the task put to a painter in asking him to paint a new picture over an old one on brittle canvas, but still to let some of the original colors shine through. No doubt, it would be quite utopian, and indeed violating our bid for ideological realism, to nurture the illusion that the educational process could accomplish that overpainting with a few bold and broad brush strokes. Educational philosophy is neither unified nor influential enough to tackle such a major assignment. However, on a less ambitious scale, step by step, the objective can be pursued with some promise of success. For, what is the substance of education if not the cumulative distillate from innumerable small lessons of experience? Therefore, if one learns to distill off and keep remembering from all the little acts and observations their common essence, rather than engraving in one's mind the more conspicuous, but trifling distractions, such habit of concentrating on essentials ought to go far in giving man a sounder rationale for his thinking and actions. The road is long and the task arduous, but if this common objective to become civilized is borne in mind as guidepost for every single step, however small, the countless component steps should add up to a resultant force of reason strong enough to overcome the inertial counterpull of prejudices based on man's more primitive *biological* instincts.

This book offers a modest effort in that direction. It aims to present sample exercises to test whether science, with its rigorous methodology, combined with logic, might not perhaps be more incisive and convincing in illustrating the middle-of-the-road way to harmony than have been the compromises of legal conciliation, economic settlement, or political accommodation. Science has had a good record of success in resolving tenacious sham controversies by proving opposing tenets to be not mutually exclusive, but rather validly coexisting alternatives. Scientific history abounds with scientific verdicts in which, on unassailably "objective" evidence, cases of supposedly irreconcilable contradictoriness were adjudicated by showing the conclusions of both contenders to have been valid. The complementarity principle of Bohr, affirming the right of coexistence of both a corpuscular and a wave concept of light; the duplicity theory of von Kries, establishing that both of two theories concerning the function of the retinal elements in color vision, formerly thought to be in conflict, were correct; the perennial fight between the embryological credos of preformation versus epigenesis—whether the whole array of organs of an adult organism is preformed as such in the egg in miniature form or whether all development is *de novo* creation—ending in the realization that there is some truth to both concepts; all these are classical illustrations of the incisive resolving power of the mature and disciplined scientific spirit, with its distaste for dissonance. On these grounds, science can truly convey an important educational lesson on how to resolve, or at least depolarize, antagonisms.

But then, science is carried on and taught by scientists, and scientists are people with all the attributes of ordinary human beings, although perhaps in slightly different proportions. As such, and while they are preoccupied with their individual pursuits in library or workshop, they need not be expected to be any more broad-minded and conciliation-bound than men in other walks of life. Indeed, violent dissent among scientists, ending in dead-end polemics, is not uncommon. However, as soon as the scientist gets out of his specialist groove and becomes mindful of his obligations to science as a whole, his sectarian acquiescence in unresolved conflict is overridden by an earnest universalist urge to strive for consonance.

The book has been conceived—and should be read—in this spirit of depolarization, of harmonizing conflicting doctrines. As the problems to be dealt with reach far beyond the scope of scientific subject matter into the area of general human concern, the treatment and conclusions of this scientific sample exercise might well lend themselves to some broader cultural extrapolations. The theme is *the recognition and scientific validation of the rule of order that pervades the universe* and culminates in human understanding.

In an earlier essay, I have tried to mediate—or, at least, moderate—the age-old confrontation of the philosophical doctrines of "holism" versus "reductionism", each vested by its adherent sect with the master key to the explanation and understanding of nature. In another, *The Living System: Determinism Stratified*, I have dealt, by way of example, with a specific scientific issue, namely, the problem of "genetic determinism", as it is understood and, more often, misunderstood. I use it to illustrate how such misunderstandings and overstatements, including some ingenious conjectures about the origin of life, can bring a sound body of scientific data dangerously close to the point at which solid knowledge turns soft and becomes warped into misshapen concepts of life and mind and of man's opportunities for self-development. In this book, the broader meaning of "determinacy" is subject to a critical reexamination, which leads to the replacement of the rigid micromechanistic ("atomistic") thinking of old by a hierarchical concept of natural "systems", in which order in the gross goes hand in hand with freedom in the small; the evidence gathered here depolarizes and, hopefully, defuses the pernicious categorical antithesis of order *versus* freedom, which has been the source of so much human strife. May strife be superseded by striving.

In sum, my leading motif in this book has been to substantiate concisely the scientific rationale and validation of civilized man's obligation to strive for a realistic and balanced perspective, in which he recognizes ideological extremes for what they are: artificially disconnected opposite ends of continuous scales of intergrading values, just fortified in their positions of antagonistic isolation by verbal symbolism and the instinctual vestiges in man of his biological past. The example of the reconciliation of the rule of order in nature with the legitimacy of freedom and diversity serves merely as a model in point. It, so to speak, sets the tune. If education would pick up the tune and amplify it, and if it were to find the proper resonance in men's minds, it would transmit to man a cultural gift from science.

> As but a line parts left from right
> And shades of gray link black to white
> And stillness waxes into noise
> And effervescence wanes to poise,
> So most extremes can be connected
> And man's contrariness corrected.

THE ORDER OF LIFE

PROLOGUE: AND WHAT IS LIFE?

Nearly a quarter of a century ago, Erwin Schrödinger, a great physicist, as inquisitive of mind as he was broad of vision, wrote a remarkable essay entitled WHAT IS LIFE? Note that the title contains a question; not an answer with an air of finality—something like "Life is. . . .", as a less circumspect man would have phrased it. In offering in the following pages another essay on the topic, I shall retreat even one step further from the pretense of answering the question; for, in the harsh light and sharpened focus of analytical examination, the very question itself, instead of coming close to being answered, dissolves. This anticipatory statement of my conclusion plainly discounts the mounting bombardment of the public mind by prophecies that science has, or is about to have, found the key to "explaining life fully in terms of the blind play of molecules", and indeed, "artificially creating life *de novo* in test tubes from purely inorganic matter."

Any such prophetic claims hinge on the premise that the question of "what is life?" has been reasonably answered to general satisfaction. That premise is unfounded, for there is certainly no unanimity on what the term "life", in that generality, connotes. To be sure, a great many propositions on that subject have been propounded off and on during the past, somewhat haltingly by scientists, more boldly by philosophers and jurists, and most assertively by theologians. But, by their very plurality and disparateness, those diverse notions have been self-defeating. They have failed to satisfy man's instinctive realization of the oneness of life and his desire for universality in identifying himself with some sort of unitary "life principle", from which he could derive the bearings for the place in nature of living creatures, himself included. Instead of the single and firm answer which he expected, he was presented with an assortment of choices, often incongruous and confusing; at times, the choice between scientific dictums and faith. And even scientific dictums on "life", on which one might choose to build one's faith, were in themselves ambiguous, being based only in part on solid evidence, and for the rest on interpretative contentions, reflecting opinions rather than facts.

Confining oneself specifically to those areas of science in whose province lies the study of the phenomena of life—the life sciences proper—one evidently is prompted to ask then why they have not yet succeeded in coming forth with a unified concept of "life". The answer, as I shall try to show, lies not so much in our still appallingly inadequate knowledge of living processes, but, on the contrary, in the fact that our knowledge of them has grown in sophistication; that is, we have come to realize that, viewed scientifically, "life" appears no longer as the simple, single, elementary entity of old which a more primitive imagination had construed it to be. If once "life" was considered a sort of essence which one might hope some day concretely to distill, purify, identify, and thus define, it now appears as but a cover name standing for the *minimum common denominator of the processes distinctive of living systems.* As we shall see later, this wholly noncommittal, "objective", description would have to be significantly supplemented if the remarkable rule of integrative order prevailing in those processes were given proper weight. Even then, however, the only cogent answer to the question of "what is life?" which life science would have to offer is: *Life is an abstraction;* a man-made noun with no well-delineated substantive meaning; a short-hand reference symbol; barely a concept. With this my essay will deal.

To many, such sober scientific "objectivity" of expression would seem quite disillusioning; to some, utterly revolting. For, they will argue, is not the sharpness of the line dividing life from death sufficient evidence for the existence in reality of such a unitary and cohesive entity that man has come to signify and dignify by the term "life"? There is no need here for engaging in a lengthy discussion of the concepts of "existence" and "reality", nor in a scientific demonstration that the boundary between life and death is biologically by no means as sharp as legal or religious norms would set it; for any counterarguments would be pointless. They would miss the point that the term "life" has truly a *dual* meaning, depending on whether we refer to our introspective experience with our own *inner* life or to our observations of the behavior of living systems in the *outer* world.

This distinction is crucial. Only the latter reference system lies truly within the scope of science, since it admits of rigorous propositions of universal and testable validity; whereas the former, the realm of our personal inner world of thoughts and dreams, ideals and beliefs, is largely private, incapable of articulation and communication to others in signs, symbols, or terms of cogently defined and unequivocal meanings. For instance, I do not know whether your and my personal sensations, when we see the same red object or feel the same cold water, are at all identical. They might differ vastly among individuals. But by referring them to a common "objective" measuring standard—wavelength for "red" and height of a mercury column for "cold"—we establish a sort of common medium for comparison and communication. And only that portion of our image of ourselves and of the universe which can be reduced to such universally communicable and comprehensible terms forms the legitimate precinct of science. The rest lies in the private domain of the individual mind.

The border between the two domains—"life" as an inner experience and "life" as an object of our experience with the outer world—is fuzzy. The content of the former—impressions, value judgements, opinions, sentiments, creative urges, moral decisions, and what not—is slowly invaded or eroded from its fringes by the rigorous methodology of science, which makes the border region of overlap disputed ground. But no matter how far and fast scientific knowledge will expand, the border, although shifting, cannot be expected ever to vanish: invasion by science of the realm of the intuitive does not portend eventual full conquest. Contrary predictions from respectable scientific quarters to the effect that scientific knowledge will succeed in engulfing totally that private sector which has appropriately been called "personal knowledge" by Polanyi may be respected, but need not be adopted, for they are, at any event, gratuitous. In fact, being sheer expressions of opinions, untestable and unprovable, based on belief rather than on evidence, they are themselves creations of that subjective, private, extrascientific realm of the mind, the primacy of which they aim to dismiss.

There is no use in carrying this discussion further. For the time being, at any rate, "life", as we know it subjectively from the primary knowledge of ourselves, is still far richer than the objective image of life which science reconstructs secondarily from the study of living beings. We, therefore, must abide by their distinctiveness. Depending on temperament, cultural tradition, and social conditioning, some people will draw comfort from this affirmation of a privileged preserve of private inner life, while others will resent or outright reject the implication of uncertainty and equivocation inherent in any picture of life refractory to strictly scientific description. The attitude for us to take here must be to leave them both to their predilections and prejudices and confine ourselves deliberately to dealing solely with that narrower domain which both of them jointly acknowledge—that limited sector of "life" amenable to objective study: the object of the life sciences. Accordingly, my discussion will focus mainly on "life" in the narrow sense of a strictly scientific issue, and it is with this restricted meaning that the term should be understood whenever it will occur. I hope that, having made this clear, I leave no room for fantastic expectations of magic formulae for "life elixirs".

To whom, then, is this piece addressed? Its aim is dual. For one thing, its general conclusions are of concern to that segment of the non-scientific public which is curious about the verdicts of the sciences in respect to life; yet right along, it also carries a major message to scientists, particularly those in various specialties, whose sectorial preoccupations reduce their opportunities for acquiring a catholic and balanced view of biological phenomena and problems. The reason for this dual aim is explained in the following:

LIFE SCIENCE: PUBLIC SERVANT

The life sciences are in a period of unprecedented drive and progress. It coincides with the awakening of man through the combined effects of self-concern, education, and mass communication to the realization that, whatever else he be, he basically is a biological being, and as such is entitled to drawing his share of benefits, both practical and conceptual, from those spectacular biological advances. His expectations, stepped up by the growing support of science from public funds, and coupled with the demonstrated potential of science for tangible deliveries, confront the man of science with a charge of social responsibilities and duties new to him and unaccustomed, which he must face and meet. He need not tear down his traditional ivory tower of research and meditation—indispensable for creative work—but he must let it become more intimately joined to the larger structure of the social complex with which science is interwoven and which, besides research, includes education and public enlightenment. Ostensibly, with the evolution of science from an affair of solitary little workshops to the dimensions of a gigantic industry, the man of science has turned into a public figure, expected to act like one.

Already he is becoming mindful of his public image, and rightly so, for it affects the measure of credence placed in his dictums by the lay public. His heightened sense of social accountability imposes on him also the obligation for exercising great care and caution in his utterances lest they be misconstrued and misapplied by an innocent and uncritical public, holding in its numbers an avalanche-like amplification factor for the propagation of error as much as of truth. His statements stay less and less confined within the precincts of science and spill over more and more boundlessly into the public scene, where they exert a growing impact on the attitudes and the behavior of people for better or for worse.

Alerted, scientists will guard against the risks. After all, they have taken up the fight against pollution—of water and the atmosphere—for which science had been blamed in the first place. Muddled language is no less hazardous, but it is just as readily combatable, once it is recognized. Not having wholly escaped Stuart Chase's "tyranny of words" with its toll of futile argument and misspent effort from confused language, the life sciences are surely prone to rise to the need for keeping the conceptual atmosphere, in which they operate, clean, and the verbal wrappings, in which they market their products, uncontaminated. *Pari passu*, the consumer, both in science itself and among the public, will learn to look for the substance behind a name, the meaning behind data, the content inside the container. The main purpose of this book is to illustrate by a sample exercise the kind of practical prevention and therapy which I believe is indicated and feasible. I am choosing for this exercise the problem commonly labeled "genetic determination", with its corollary, the "origin of life", but other issues of equal concern to modern man could have served as well (e.g., brain function and the brain-mind problem, or the ecological continuum of body and environment).

In undertaking this task, I have been keenly aware of its inherent, almost prohibitive, difficulties and of my limitations in surmounting them. For one thing, the task called for a critical examination of the practices through which conclusions in the life sciences are reached and communicated—a sort of intellectual audit of the conceptual operations involved—entailing unavoidably an entanglement in age-old philosophical controversies, which better men have been unable to resolve. I partly tried to overcome this hurdle in a preceding essay, listed as (B) under "References", but still have to face it head-on in the present book when tackling the issue of "determinism", clarification of which is crucial for the understanding of what "genetic determination", and in a derived sense, "individual freedom", really mean. At any rate, an auditing job requires the impartiality of the outsider, and so I had to place myself temporarily beyond the pale of loyalty to any particular group, scientific or other. It is not pleasant for a doctor to diagnose an illness in a friend, but if he can follow through with sound advice on cure or prevention, he feels comforted. If some of my diagnostic remarks were to vex some members either of the scientific fraternity or of the educated public for seemingly insinuating callousness or gullibility, respectively, let me point out that statements of trends and averages for groups never apply to individual members of the group. Evidently, when we speak of bringing science and the public closer together, we imply that they have not yet been close enough. It therefore is a logical necessity to identify what has kept them apart, and to do this has been the sole intent of pointing out the difference in their positions.

The most formidable obstacle, however, has been presented by the attempt to carry out this two-front mission in a single act, with the two fronts as disparate as science and the public. The fact that I had to place myself in the middle between the professional and the layman in executing what essentially amounts to a scientific and epistemological exercise in terms as rigorous as the scientific code dictates, and at the same time inviting the lay public as an audience, has made me incur the serious risk of failing both sides. I have tried to present the strictly scientific documentation, which had to be adduced in many places in order to make a point, without recourse to technical terminology and without assuming any but elementary scientific knowledge. Nevertheless, the product of the amalgamation of the missions has still some of the aspects of a disproportioned hybrid. Hopefully, it will, these formal defects notwithstanding, not be a sterile one, but on the contrary, possess the genetic virtue of hybrid vigor.

A VERBAL TRAP: "GENETIC DETERMINISM"

Research is science's tool to provide the information on nature which the human mind then interprets, orders, and fashions into scientific knowledge in the form of formulae, theories, principles, and concepts. To serve their scientific purpose of "objectivity", these formulations ought to be, as I remarked, couched in unambiguously defined terms, which leave no room for the arbitrariness of subjective interpretation. For scientific propositions and conclusions to be generally communicable and comprehensible, the symbols in which they are articulated, whether words or figural signs, must convey standard meanings. Absolute attainment of this ideal is still rare, but the degree of approximation to it is a measure of the maturity of a science. On this score, the physical sciences are, by and large, farther advanced than the life sciences; but, then, the phenomena and problems in the latter are also enormously more complex.

The designations of a physical phenomenon, such as crystallization or electromagnetic induction, carry a far more sharply defined meaning than, for instance, a biological term like "growth", which may refer promiscuously to such different kinds of incremental phenomena as the assimilation of food, cell multiplication, increase in length, protein synthesis, or just mounting complexity. A given species of molecule can be rather well described and identified in terms of its chemical and physical properties, but when the biologist calls it a "hormone", he no longer refers to criteria of the molecule itself, but he sums up a functional interpretation by the observer: the messenger role which he has observed the molecule to play in the household of the organism; to wit, the relation and service of the molecule to other entities. While useful for purposes of general classification, the term thus is totally unfit for diagnostic identification, and, in fact, the most disparate chemicals are being employed by organisms to serve as "hormones".

The list of similar instances of latitude implicit in biological terms could be widely expanded. The margin of terminological indefiniteness which they illustrate is partly, as I have indicated and shall further elaborate later on, an expression of the enormous complexity, variability, and actual indeterminacy inherent in biological phenomena. In part, it also reflects the relatively young stage of the life sciences, in which new phenomena and new principles turn up at such a fast rate that it would be utterly impossible to follow each through to a satisfactory completion. Those left behind in their unfinished state are usually given provisional labels, hopefully to be picked up at some future time for further study and elaboration.

It is in this latter class that misconceptions are apt to arise if the provisional character of the label is not duly stressed; if it is not explicitly presented as a cover term for a topic in need of further exploration or conclusive verification, but intimates that the desired knowledge and conclusiveness are already at hand. Since suggestive terms have a way of guiding, and by stereotyped recurrence grooving, habits of thought, the imperceptible change in usage of a term from a temporary stop-gap designation of anticipated knowledge to a permanent surrogate for actual knowledge bears critical watching. Periodic tests of the state of congruity between the factual content of a phenomenon (or principle) and the identifying "brand" name under which it runs can go far in checking incipient discrepancies. Such intellectual check-ups follow the maxim of public health, according to which "an ounce of prevention is better than a pound of cure". This is the spirit in which I shall approach the following critical review of the meaning of the term "genetic determination"—as a sort of clinical symptomatology.

Let me start from the familiar vernacular, which offers us a telling diagnostic clue. It lies in the common habit of referring (in popular and semiscientific, but also some professional, writings) to "characters" of living organisms as "genetically determined". Whether applied to structures, functions, mental traits, or behavior patterns; to embryonic stages, adult life, or senility; to prowess or debility; the term, if taken literally, is factually inaccurate. If it is used simply in a vague metaphoric sense, it is ambiguous, hence, scientifically meaningless, and in its connotation of rigid predetermination of the true fate of a living system, deceptive. For if the term "determined" is interpreted to mean rigid fixation of a course of events from beginning to end, without latitude for excursions, indeterminacy of details, and adaptability to the contingencies to be met on the way—and in the case of man, at least, freedom of choice—man would be led to accepting all of his life as preordained, all of his striving as futile, all of his hopes and aspirations dashed by a scientific verdict of abdication to unmitigated fatalism.

And the term "determine" means just that; witness some of its definitions in standard dictionaries as "to lay down decisively or authoritatively"; "to fix beforehand; to ordain, decree"; "to fix or decide causally"; and so forth. Either this is the way the "determinative" action of genes is construed; or if it is not, then the term should be purged from the scientific language and expunged from public use, just as a mislabeled drug containing poison is recalled from the market under the Pure Food and Drug acts, which require that all items offered for public consumption as food or drugs list on the labels the detailed content of ingredients.

What is misleading in the term "genetic determination" is that it conveys the notion that the development of an organism is simply the mechanical product of a bundle of linear "cause-effect" chain reactions, reeling off in rigid sequence according to a minutely predesigned plan of clockwork precision. That notion, reinforced by the anthropomorphic language that endows genes with the powers of "dictation" and "control", rests on a basic misconception of the nature of biological processes in general and of developmental dynamics in particular. Scientists familiar with the facts, of course, know better. They do acknowledge the wide variation left to an organism in executing the "genetic blueprint" encoded in the germ cell, and they express that knowledge in the distinction between "genotype" and "phenotype"; the former referring to the primordial genic endowment of a given individual; the latter, to what the individual actually turns out to be. The distinction used to carry an undertone of strictly preformationist thinking, as if the "genotype" were the true ideal type in the Platonic sense, while the "phenotype" represented its corrupted realization. Though mitigated by the designation of the postgerminal course of development as "epigenetic", some residue of the old preformistic purism has lingered on, and the description of any terminal "character" of an organism as "genetically determined" is clearly a relic of that old tradition.

THE ORGANISM: A SET OF CHINESE BOXES

To lift an atavistic stereotype out of its groove calls for more than just opposing one fiat by its anti-fiat. It calls for a sober examination of the factual record. And indeed, the record reveals the basic fallacy. It lies in the presumption that the organism is a sort of chemical slot machine. If recourse to

The Order of Life

machine analogy is at all necessary, one might as well point to the many features in living systems that have rather the aspects of pinball-machines, in which a succession of identical events, started from exactly the same point under identical conditions, end up not at the same destination, but in the broad statistical dispersion of a Gaussian population, with no individual course precisely predictable from the start. Ever since 1925,[1] I have presented the case for the dynamics of organisms as being in the nature of *systems*, rather than machine operations, but even though the distinction is crucial in the present context, the best I can do here is confront the abstract literary notion of "gene-determined" events with a more concrete picture of how developments "from gene to character" proceed in actuality. For brevity, let me refer to Figure 1.

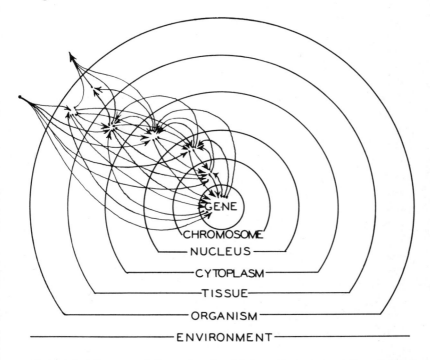

Figure 1. Diagram of the network of interactions among the sub- and subsubsystems in the hierarchical organization of a living organism.

Any living system can be represented as a hierarchy of concentric shells of subsystems, each shell differing from its outer and inner neighbor by some distinctive properties (e.g., composition and distribution of content, physical consistency, reactivity, etc.). As indicated in the picture for a higher organism, the genes are embedded in the chromosomes, the chromosomes lie in the nucleus, the nucleus is surrounded by cytoplasm (which furthermore contains organelles), the cells are incorporated in the tissue matrix, bathed by the "internal milieu" of blood and lymph, and the whole body then faces the outer environment. For each shell, the adjacent ones are evidently "environment" in which they lie entrapped. None of the shells can ever be truly "isolated". All one can do is change them from one environment to another, however drastic the change may be (as in the introduction of a mutant gene into a chromosome, the transplantation of a nucleus into strange cytoplasm, the transfer of a cell from the body into tissue culture, or the move of an individual into a different cultural climate). No part of this complex fabric can, therefore, ever be singled out and treated as if it were alone. Each can be described and understood solely in reference and in relation to its particular immediate environment with which it is in dynamic interaction. Since each part, in turn, constitutes environment for its neighbors, the true dynamics of an organism must be envisaged as the intricate web of interactions symbolized by the arrows in the diagram.

By the same token, these arrows represent the only channels of communication among the different layers, inward and outward. No agent or signal, except perhaps radiation, in passing from outer to inner shells or the reverse can escape the screening and potentially transforming effects of the intervening zones. This is as true of hormones injected into a tissue in their eventual impact on genes, as it is, in the reverse sense, for the product of a genic interaction in the innermost sanctum traversing the more outer strata, where it comes to our attention. This fact ought to dispose of the glib presumption that any agent we inject into a body gets to the site of genes unmodified and that, conversely, what we record as a "gene effect", is really the primary and direct manifestation

The Order of Life

of a "gene product", rather than a derivative of many steps of intermediate transactions in passage from the inside out. There can be no prediction *a priori* as to what passes without alteration and what does not. It is a matter for methodical research to replace the arrow symbols of our diagram by concrete information; but to ignore them, erase them mentally, or just give them names, will certainly not do.

THE EGG: A PATTERNED ORGANISM

Let us now look at development, embryonic as well as postnatal, in this light. An egg is a mononuclear uncellulated organism.[2] Its cortex is an orderly mosaic pattern of cytoplasmic domains of disparate molecular composition and constitution, and consequently of different reactive properties. Therefore, as the egg divides, the multiplying nuclei come to lie in regionally differing cytoplasmic environments, to which they then react differentially. So, the chromosomes, and in further consequence, the genomes of different cleavage cells are very early confronted with a set of typical inequalities of surrounding conditions, depending on the particular sectors of the cytoplasmic map in which the nuclei happen to have landed. Their very first interactions, in modifying the properties of the respective cell territories, add further to the regional disparity among the "shells" surrounding the essentially identical genomes. The genes are thus trapped in cages, like prisoners, allowed communication with outer nonneighboring shells, such as with other cells, only through the mediation of intermediary agencies, which translate, screen, modify, or even suppress their "messages". In turn, whatever information the genes can obtain about what goes on in other parts, reaches them only in the form of commuted signals that have filtered through to them, the nature of the filters being crucially different in the different cytoplasmic areas— the different prisons, or to put it more pertinently, "schools", as it were.

The main thing to keep in mind is that even the ooplasm, in which the first step of cell division will take place and with which the daughter nuclei and their later descendants will have to deal, is an environment endowed with a typically differentiated pattern of its own. Experimental evidence available confines the territorial integrity of this primary space pattern to the egg "crust" (cortex) which, being an interfacial boundary zone, has the solid-state property of resisting the randomization of fluid bodies by thermal agitation.

This surface map of disparate sectors is, of course, itself the product of the developmental "pre-oval" history of the egg as a member of the community of somatic cells of the maternal body and, as such, carries the stigmas of the whole long chain of interactions dating back to the first encounter of the parental genome with grandmother's egg pattern, which then could be referred back to the next anteceding generation, and so forth; one might want to spin out this argument into a continuous, never interrupted, chain of backwards projections, which in the version of the old-time preformationists would have started, for man, in the ovary of Eve, and in the speculations of modern evolutionary theory, in the first viable germ to have emerged from the primordial soup.

In our present context, it is wholly irrelevant whether or not we could extricate ourselves from the logical tangle to which such an *a posteriori* argument can lead, for our concern at this point is with the dynamics of the here and now—the actual transformation of an actual egg into an actual mature individual. In this regard, the crucial thing to realize is that the ooplasm of the actual egg is decidedly not just a post-primordial soup to which the captive genome inside imparts "organization" by one-way effusion of "information". Rather is there a dual system of order, a grosser one of differential boundary conditions on the outside and the more refined one of the genic array on the inside, neither of which can "act" independently and "dictate" unilaterally; if the simile of human language is to be used at all, their interaction must be described as dialogue.

Moreover, adopting the linguistic analogy between the genetic code and a three-letter syllable and following it through the transcription to word assemblies (from DNA to

The Order of Life

RNA's) and their translation to isomorphic sequences of amino acids composing proteins, one still is left with the vexing question: "What sort of a principle is to account for the syntax?" Why is it that at no stage of its development an embryo is simply a bag filled with a random scramble of proteins, whether structural or enzymatic, "acting" on lesser molecules equally scrambled? Whence stem its features of organization that, in our metaphor, are comparable to the structure of whole sentences and paragraphs, which still make sense even though the sequential order of their constituent words, themselves replaceable by synonyms, is anything but stereotyped?

These questions, if asked at all, are mostly brushed aside by two catchwords: "information" and "control". The genome is supposed to provide the extragenic system with the "information", which then in turn would feed back to the very informants the cues for group organization, which they did either not possess or not understand in the first place. This is the kind of logic that is epitomized by the proverbial feat of cheating gravitation by lifting yourself up by your bootstraps. It is like proposing that a number of separate magnetic needles could ever orient themselves in a common direction without the guidance of an outer magnetic field. Could a random mixture of even such uniquely structured elements as nuclear constituents ever "instruct" an environment, if it were equally randomly dispersed, to generate the typical developmental space patterns observed? Without dwelling further on this brand of logic, let me just caution innocent readers of recent literature to be on guard against the scurrilously perverted usage of the term "information" as a cover for "ignorance", to wit, "lack of information"; reminiscent of the Latin jibe of *lucus a non lucendo*.

The term "control", in the sense of "command", is not much more meaningful. It refers correctly to a basic feature that distinguishes organized behavior from sheer randomness, namely, some orderly restriction of the degrees of freedom of the studied system. But to envisage how to get from this negative feature to the positive laying down of a definite space pattern, unless there were a pre-existing spatial pattern of the "controlling" agency itself, would evidently again require quite a slight of logic. Actually, no realistic proposal for the *de novo* emergence of space patterns in random molecular populations by mere "control processes" has ever been advanced. These remarks are in no way directed against the well documented and fundamental role of "self-control" of biochemical dynamics and kinetics by feedback principles. They merely caution against the unwarranted verbal extension of those principles beyond their legitimate range of applicability. They are definitely inapplicable to the problem of primary space patterns. I am basing this conclusion on the fact that although having myself been the author of a rigorous mathematical model of molecular "feedback control" of growth,[3] I have also had plenty of firsthand experience with the actual process of morphogenesis to know that space patterns do not arise that way.

To sum it up, if the existence of some spatial organization in the ooplasm had not been actually verified, as it has, one would have had to postulate it as the only rational alternative to the invocation of some special "organizing principle" in the sense of sundry vitalistic doctrines. At any rate, the fact that from the very start of embryonic development the genome is in interplay with a spatially and chemically patterned outer system deprives the term "genetic control" of development of its monopolistic ring. There is a "plasmatic control" as well. Lest this be misconstrued to mean merely the presence in the ooplasm of free-floating particles with genelike coded "information", which have been amply demonstrated (ribosomes, plasmagenes, etc.), let me stress that, being freely mobile, these could no more account for primary overall space patterns, including their own differential localization, if any, than could the genes locked inside the nucleus. The prelocalization of chemically and structurally disparate sectors in the ooplasmic matrix is thus not only a demonstrated fact but a logical postulate. The recognition of this duality of organized counterpoles in mutual interaction—the genome innermost and the ööplasmic map outermost—thus returns the genome from its semantically appropriated status of an "operator" to its true role as a "cooperator". There is no step in development—or any process in a living system, for that matter—in which the genetic background is not integrally involved, but there is no phenomenon or attribute of life, either, in which the genome would be solely instrumental as a kind of omnipotent autonomous ruler, in which capacity it has often been presented explicitly or by indirection.[4]

In a wider sense, this rectification of a one-sided perspective of "gene function" has implications way beyond the problems of development, as it bears also on our conceptions of brain function, behavior, human relations, and, generally, the problem of "determinism". But let us, for the moment, put this aside and return to the embryonic germ and the "determinism" inherent in the fact that it will give rise to an essentially predictable result. The salient question is: "How rigorously predictable? Down to the most minute degree of microprecision, or just within a given overall canon, allowing for so much latitude of the detailed internal events that their individual uniqueness makes them unpredictable? And in the latter case, what is the margin within which certainty of detailed prediction must abdicate to a demonstrable uncertainty principle?

This question, as one notes, would take us right into the age-old philosophical arguments about "determinism", obviously crucial to our discussion, although here we deal with the far more restricted empirical problem of the "determination", genetic or otherwise, of developmental events. It so happens, however, that from this narrower base we have gleaned a lesson that has a fundamental bearing on those broader issues. For we have learned that, speaking concretely, the very antithesis "determinate vs. indeterminate" is spurious. A macroevent can be fully determinate in the sense that, given the premises, we can predict the general outcome, the confidence of our prediction resting on the infallibility of countless earlier experiences; yet at the same time, the component microevents involved might take courses that are in detail absolutely unpredictable and unique—unpredictable, because being unique and nonrecurrent, they have had no exact precedent in our experience. Let me, however, at this point, first clarify my position with regard to the problem of "determinism" in general, which I have exposed more elaborately in the two essays listed in the References as (A) and (B).

DETERMINISM STRATIFIED

To me, the main flaw in judging "determinism" lies in its equating science with the doctrine of microprecise causality, or, as I shall call it in the following, *microdeterminism*. This brings me to the major lesson to which I have been building up in this article and which I have anticipated in the term of "Stratified Determinism", which is precisely what the study of living nature teaches us.

To judge whether or not there are philosophers or theologians who might get comfort from a scientific image of the universe made up of a mosaic of discrete particles operating by laws of microcausality is beyond my ken. All I submit is that modern science cannot deliver such a picture in good faith, least of all life science; and since all science is the product of human brains and brains are living systems, it is quite likely that this abrogation of scientific rationale for microcausality applies to science in general.

Scientifically, the term "determined" can only mean "determinable"; and similarly, "indeterminacy", whether in the sense of Heisenberg or in the way I shall use the term, can only mean "indeterminability". The scientific concept of "determinability" is of decidedly empirical origin. As we observe a given macrosample of the universe over a given stretch of time, we note certain unequivocal correlations between the configurations of its content at the beginning and at the end of the period of change. If then we find those correlations verified with recurrent consistency, we set them up as laws, from which to extrapolate future changes with a sense of certainty. As our primary experience in this operation has only correlated macrosamples with macrosamples, predictability based on it can likewise be no finer than macroscopic. Therefore, legitimately we could only speak of "macrodeterminacy".

The concept of "microdeterminacy" is then derived sec-

The Order of Life

ondarily from a hypothetical downward extension—atomization, as it were—of empirical "macrodeterminacy". That hypothesis submits that one would observe the same high degree of consistency of correlation from beginning to end that had been ascertained for the macrosample to hold true for every one of its fractional samples. In other words, the structure of the well-defined macrochange would be simply a composite of the mosaic of microchanges assumed to be equally well-defined, even if not necessarily determinable.

This tenet is demonstrably untenable in its application to living systems. We have recognized the states and changes of such systems as being conservatively invariant over a given period, and hence predictable, without a correspondingly invariant micromosaic of the component processes. We have to conclude, therefore, that the patterned structure of the dynamics of the system as a whole coordinates the activities of the constituents. In atomistic microdeterministic terms, this coordination would have to be expressed as follows: Since any movement or other change of any part of the system deforms the structure of the whole complex, the fact that the system as a whole tends to retain its integral configuration implies that every change of any one part affects the interactions among the rest of the population in such a way as to yield a net countervailing resultant; and this for every single part. Couched in anthropomorphic language, this would signify that at all times every part "knows" the stations and activities of every other part and "responds" to any excursions and disturbances of the collective equilibrium as if it also "knew" just precisely how best to maintain the integrity of the whole system in concert with the other constituents. Although rarely expressed so bluntly, much of this imagery lurks behind such equally anthropomorphic terms as "organizers", "regulators", "control mechanisms", and the like, which particularists have had to invoke in order to fill the information gap between what one can learn from isolated elements and a valid description of group behavior.

The Boltzmann theorem and thermodynamics have realistically bypassed that gap by confining safe statements about macrorelations to macrosamples only. They relate unequivocally the average state of a system at time t_1 to its average state at time t_2, but realize that tracing an individual molecule through that course is not only unfeasible but would be scientifically uninteresting and inconsequential: it would in each individual instance and instant be of nonrecurrent uniqueness, hence valueless for any detailed predictability of future microevents. If physics has had the sense of realism to divorce itself from microdeterminism on the molecular level, there seems to be no reason why the life sciences, faced with the fundamental similitude between the arguments for the renunciation of molecular microdeterminacy in both thermodynamics and systems dynamics, should not follow suit and adopt macrodeterminacy regardless of whether or not the behavior of a system as a whole is reducible to a stereotyped performance by a fixed array of pre-programmed microrobots. Since experience has positively shown such unequivocal macrorelations to exist on various supramolecular levels of the hierarchy of living processes in the documented absence of componental microdeterminacy, we evidently must let such positive scientific insights prevail over sheer conjectures and preconceptions, however cherished and ingrained in our traditional thinking they may be.

MACRODETERMINACY

In order to drive home this lesson, I am adding in Figure 2 a diagrammatic model of macrodeterminacy. This primitive diagram shows the transition of a living system from a state S' to a state S''. As is indicated in Figure 2a, the system S', comprising subsystems A', B', C', D', E', develops between times t_1 and t_2 into the modified system, S'', each subsystem of which at t_2 can be traced back to a corresponding subsystem at t_1; this makes the pattern of t_2 explicable, that is, predictable, hence, determinable or "determined", as a direct transform, piece by piece, of the macroconstellation of component subsystems at t_1, the subsystems having kept their relative positions, i.e., spatial relations.

The Order of Life

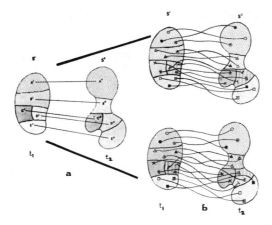

Figure 2. Diagram of development of an embryo at stage S′ to stage S″ (between times t_1 and t_2) with topographic correspondence of macrodeterminacy (at left) without detailed microdeterminacy (at right).

This kind of mosaic correlation between two stages permits an embryologist, for instance, to identify specific regions of an early embryo as being the predictably earmarked forerunners for the formation of heart, liver, kidney, brain, etc., respectively. Yet in looking at the diagram 2b, we note that such clear-cut correlations no longer hold for smaller samples of each subsystem, represented by the various symbols. In other words, if we were to follow individual cells of these various prospective organ areas from the earlier into the later period, we would find them to take far more fortuitous courses, differing individually from case to case; this is indicated by the lack of correspondence between the upper and lower sets of lines connecting symbols from S' to S" for two embryos of the same species. This fact is so well established in embryology that one has even gotten to referring to the process which changes a cell from its originally "indeterminate" and multivalent condition into one definitely committed for a given fate, as the process of "determination".

I could go on to confirm the validity of this principle of "determinacy in the gross despite demonstrable indeterminacy in the small" for practically any level and area of the life sciences. In order to take account of this hierarchical repetitiveness, I have suggested the simile of "grain size" of determinacy as an empirical measurement for the degree of definition and predictability at any given level. The mosaic of organ rudiments mapped out in the early embryo, for instance, is very "coarse-grained", whereas the mosaic of genes in the chromosome is far more "fine-grained" (see below).

One could easily turn this renunciation of the sovereignty of microdeterminacy into a positive scientific declaration in favor of the existence of free will. I prefer to give it a more restrained interpretation, for it really implies no positive commitment. What it does is simply remove the spurious objections and injunctions against the scientific legitimacy of the concept of freedom of decision that have been raised from within the scientific sector or from other camps leaning on supposedly scientific verdicts. I cannot see that science can prove free will but, on the other hand, I can see nothing in what we know in the life sciences that would contradict it on scientific grounds. To go beyond this neutral statement would be a matter solely of private belief, conviction, or opinion, without objective substantiation.

A CANON FOR DETERMINACY

Lest there be misunderstandings about my thesis of macrodeterminacy which is not explainable in terms of aggregates of microdeterminate events, let me review briefly the major way stations to this conclusion.

1. Nature presents itself to us primarily as a continuum.
2. In scanning this continuum, we recognize complexes of phenomena which retain identity and show a high degree of stability and persistence of pattern, in contrast to other samples with less cohesive features.
3. Success of science over the ages has validated the abstraction involved in our dealing with such reasonably constant entities as if they had an autonomous existence of their own.

The Order of Life

4. Some phenomena of nature can be reconstructed in practice, or at least in our minds, from the analytical knowledge of properties, interrelations, and interactions of such putatively isolated entities.
5. Some of the sciences, particularly the physical sciences, have confined themselves mostly to the consideration of phenomena amenable to the treatment according to 4.
6. In the life sciences, there are likewise many questions which can be answered satisfactorily by the recombinatory method according to 4.
7. Understanding of the integrality of a living system, however, has proved, on logical grounds, refractory to the same methods, and the empirical study of life processes has discounted, on factual grounds, the probability of future success.
8. The preceding point implies that it is logically and factually gratuitous to postulate that the methods of 4, successful in 5 and 6, must necessarily also be sufficient to restore completely the loss of information about the dynamics of *systems* suffered in their analytical atomization. By the same token, restrictive injunctions against descriptions of living phenomena in terms other than those compatible with 5 and 6 can no longer be upheld.
9. The systems concept proves applicable to the description of those phenomena in living systems which defy descriptions purely in terms of micromechanical cause-effect chain reactions; it thus lends substance to the principle of systemic organization.
10. Applying the systems concept, an organism as a system reveals itself as encompassing and operating through the agency of subsystems, each of which, in turn, contains and operates through groups of systems of still lower order, and so on down through molecules into the atomic and subatomic range.
11. The fact that the top-level operations of the organism thus are neither structurally nor functionally referable

to direct liaison with the processes on the molecular level in a steady continuous gradation, but are relayed stepwise from higher levels of determinacy (or "certainty of determinability") through intermediate layers of greater freedom or variance (or "uncertainty of determinability") to next lower levels of again more rigorously ascertainable determinacy constitutes the principle of *hierarchical organization*.

12. Although I have emphasized for didactic reasons the relatively conservative features of systems, the unidirectional change of systems must not be overlooked. We find it expressed, for instance, in the mutability of systemic patterns in evolution, ontogeny, maturation, learning, etc., as well as in the capacity to combine systems into what then appear as super-systems with the emerging properties of *novelty* and *creativity*.

This set of twelve points represents a sort of conceptual canon, based on empirical studies, against which theoretical pronouncements and formulations in the life sciences ought to be checked. As you will recognize, some statements in the current literature would fail the test of validation in terms of these criteria, while some others would even seem totally irreconcilable with the principle of *stratified* determinism. One in this latter class that comes immediately to mind is the prevailing notion of *genetic determinism*.

DETERMINATION: MEANING WHAT?

Let me return now to the specifically developmental question, namely, how rigorously can the eventual fate of any particular component element of an embryo, organ, tissue, or cell, actually be predicted. The pertinent facts, sketched above as bearing on this question, can be summarized as follows:

Figure 2 has symbolized the fact that, while definite regions of the early embryo are predictably the rudiments for the formation of heart, liver, kidney, brain, etc., the individual fates of the cells derived from those areas are still unsettled. In other words, overall regularity in the gross is attained

The Order of Life

and maintained not as a mechanical result and a reflection of a corresponding underlying regularity of rigidly stereotyped behavior of the component elements down to the smallest detail, but on the contrary, in spite of a high degree of vagrancy among the latter. The individual component does not "know" where and in what specialty it will end up until it has been well on its way and given a defining "cue". The cue will vary in accordance with the position the part assumes in the domain of the respective subsystem; and in dynamic terms, geometric location simply denotes the constellation of forces and processes at that point. Since such force and process systems of specifiably patterned distributions are conveniently classified as "fields" with the attributes of unity and continuity, in contrast to the discontinuous packages of properties denoting discrete subunits or particles, any process of "determination" must be regarded in a dual light: as the response of discrete unit members of a group, whether molecules, genes, cells, or higher organisms, on the one hand, to the specifying conditions ("cues") of a patterned dynamic continuum ("field") of which they form the constituent parts, on the other.

The hierarchical structure of natural order, manifested in this dual aspect of "determinism", is well illustrated in the molecular ecology of the individual cell.[5] A cell retains its identity far more conservatively and remains far more similar to itself from moment to moment, as well as to any other cell of the same strain, than one could ever predict from knowing only about its inventory of molecules, macromolecules, and organelles, which is subject to incessant change, reshuffling, and milling of its population. I am referring here to the unity of the *living* cell, which unlike some current unrepresentative models, is neither a test tube mixture of compounds with unlimited collision (i.e., interaction) probabilities, nor a rigidly compartmentalized space, but is a system whose inner order stems from the fact that its component subunits are subordinated to the integrative dynamics of their collective operations.[6]

This is in essence what Figure 2b represents graphically for all orders of magnitude on the hierarchical scale. At the level of the cell, the symbols in the diagram (squares, crosses, circles, etc.) stand for subcellular units, while at the level of the embryo, they stand for individual cells. In short, returning to our starting proposition, whereas larger domains of the early germ (e.g., presumptive ectoderm, mesoderm, entoderm) are rather definitely blocked out, endowed with distinctive properties, or what we call "determined", the individual cells that come to partake of them are not. The latter acquire their differential distinctions gradually as a result of having occupied distinctive positions within the field of their domain which provided them with local cues. At that stage, an observer can forecast from his past observations that a given group of cells will most likely end up, for instance, in the brain, rather than in the heart or liver, while at the same time he could not possibly venture a similarly definite prediction for any of the member cells of the particular group, for they themselves do not yet "know" their destinations, being still all alike. The fact that they truly are alike and not just appear to be because of lack of discriminatory resolving power of our techniques has been amply verified experimentally by shuffling them around arbitrarily and proving their equivalence and equipotentiality.

What we learn from this is that the fate of a larger biological entity—in the present case, a block of tissue in an embryo—may be firmly "determined", because we know what, on the whole, will become of it, while the precise fates of its individual components of lower order of magnitude— the cells—are still indefinite. This lesson is substantially the same as the one I intimated above, on a lower level, for the individual cell, quite apart from its membership in the multicellular community of the embryo. As stated there, the single cell, as a separate entity in its own right, lends itself, in its unitary behavior, to predictive statements of infinitely greater certainty than could be made about the more erratic behavior in space and time of its subdivisions of still smaller order. The recurrence of this experience in several orders of magnitude is of signal importance.

HIERARCHIC SYSTEMS: MEASURED BY "GRAIN SIZE"

Our reasoning, reflecting simply empirical facts, then leads to the following conclusions. Evidently, just to designate a biological system as "determined", hence its behavior as "determinate", is meaningless unless one specifies whether one means determined "as a whole" or "in its parts"; and even then the definition remains incomplete unless one can extend it to the parts of the parts and further to the parts of the parts of the parts, and so on down; in last consequence, to the subatomic level, where Heisenberg's "indeterminacy" principle presents a formal, though perhaps spurious, analogy. The crucial fact that the criteria of "determinacy" and "indeterminacy", meaning of "certainty" and "uncertainty" of predictability, as here applied, pertain to discontinuous levels of different orders of magnitude in a hierarchical sense, rather than being rangeable on a continuously graded scale, seems to call for the introduction of an appropriate descriptive term. Simply for illustrative efficacy and decidedly without any pretence at philosophical profundity, I propose to speak of the "grain size" of the determinacy of the properties of a living object. Its measure is the minimum size level down to which an organismic pattern can be dissected into a mosaic of definitely specifiable component parts, but below which regularities referable to that pattern cease.

The term is borrowed from the field of vision. In pattern vision, the retina of our eye decomposes any incident image into as many discrete fragments, or "grains", as there are visual receptor cells covered by it, each such cell reacting as a unit. In insects with compound eyes, the corresponding unit elements are the ocelli. Similarly, on a photographic plate, an object registers as a mosaic of silver grains; on a television screen, as a mosaic of "flying spots", and so forth. All those devices have in common the fact that there is a finite limit to their resolving power for features of the outer world: continuous patterns are reproduced with more or less loss of definition as more or less coarse mosaics of separate elements, the coarseness varying inversely with the respective grain size.

What appears here as limitation of reproductive technique is quite analogous to what we recognize in organisms as a basic property of living nature. A pattern has "mosaic" character if after having been broken up into fragments, it can still be reconstructed from the fragments piecemeal without any further clues, each piece evidently having had adequate inherent definition to mark indelibly its place relative to its neighbors in the total picture—comparable to the dismantling and reassembling of a machine. On the other hand, any pattern which is irretrievably lost by physical disintegration, that is, which cannot be resurrected after its dismemberment by sheer reassembly, has "systems" character; a system being a natural entity the components of which are so intimately interdependent and enmeshed in a common dynamic network that they can only be defined in reference to the configuration and overall order of the total complex and lose their identities when separated from that context (see references (A) and (B)). A mosaic is like a jigsaw puzzle; a system is like a spiderweb, in which the breaking of a single thread affects the structure of all the rest. Yet, a system can turn into a mosaic state whenever the free dynamic structure of the system state becomes, in some degree, fixed or "frozen" into a more rigidly consolidated condition. From that conversion, the former system emerges in a state of parcellation, each parcel endowed with properties for the separate elaboration, independently of the others, of the particular local fragmentary features of the total pattern from which it had been isolated. Parts that earlier had been undefined as to their future specification, hence, "undetermined" and interconvertible, like the small units in Figure 2, have since acquired definition and become "determined": in other words, coarse-grain determination has become more fine-grained.

Such a stepwise progress form systemic to mosaic states— from the more general to the more specific—has long been recognized as a basic principle of development; it used to be referred to as the trend from "dependent differentiation" to "self-differentiation".[7] In terms of present knowledge, it can be spelled out more concretely. It is exactly in those terms that the concrete record of developmental events differs radically from the idealized mental picture, evoked by current terminology, of a microprecise "genetic determinism".

Let us look at that record, starting with the egg. As I outlined before, there is in the ooplasm from the very start a topographic mosaic of differently constituted domains. There being only few of them, sometimes no more than three, that mosaic is of an extremely coarse grain, indeed. Compared with it, the grain of the encapsulated genome, as measured, let us say, in cistron units, is immensely finer. Whatever changes each domain incurs in its first interaction with the uniform gene population, one would not expect offhand that the coarseness of the grain of the mosaic pattern would have diminished: a set of three should still be a set of three, albeit altered three. This expectation is not necessarily valid, for experiments with delayed or suppressed insemination have shown that the intrinsic dynamics of the primordial domains itself can go on to produce further segregations of content prior to, and even in the absence of, the first nuclear and cytoplasmic cleavage step. But even with this qualification the mosaic is still very gross. The opportunity for its progressively finer elaboration up to the mature individual comes only secondarily with cell multiplication, which opens two novel sources for further diversification among parts of the embryo, namely, (1) the intrinsic field patterns of emerging multicellular communities, and (2) the interactions between dissimilar communities, adjacent or remote.

The internal progress of the specification of parts within an erstwhile uniform domain, which gradually transforms the latter from a systemic state into a mosaic state, has been variously referred to in the past as "segregation", "emancipation", "individuation", "autonomisation", and so forth. It is still rather poorly understood, mostly because its study has been sorely neglected. Three facts seem, in the main, to account for that neglect: (a) the problem has not been properly recognized and circumscribed; (b) its study is difficult, for being a group phenomenon, hence lost in the analytical concentration of attention on isolated fragments, it requires a methodology that goes beyond the ken of "molecular biology"; and (c) by a quirk of history, interest and research have been siphoned off from its study to that of effects exerted from without, listed above as (2) and usually called "inductions". Redress from that imbalance may be expected from the revival of interest in the "self-differentiation" of explanted embryonic fragments in "organ culture", but for the present, our knowledge in these matters is rather dim.

FIELD PATTERNS: PROPERTIES OF COLLECTIVES

In order to bring the notion of "self-organization" down to earth, let me show a bathing beach on a Sunday (Figure 3).

30   The Science of Life

Figure 3. Aerial view of bathing beach; water and tidal zone on top, access road for motor traffic at bottom (from reference (A).

Consider the people as molecules. The heavier border on top is the condensed belt of hydrophobic bodies adsorbed to the water-beach interface. The dark clusters inside the mass clearly mark domains of attractive forces, presumably emanating from sources of nutrient and stimulant attractants. Their equidistant spacing indicates mutual repulsion through forces of competition; and so forth. The analogy is not at all facetious. It cuts deep into the heart of our topic, for it exemplifies basic features of self-organizing systems. I could have gone on, for instance, to demonstrate how a random mixture of isolated single cells, obtained by dissociating an already functioning embryonic kidney, then scrambled, lumped and properly nourished, can reconstitute itself into a typical remarkably well-organized miniature kidney; or how similarly scrambled cells of embryonic chick skin in tissue culture can grow into normal feathers; all this entirely by "do-it-yourself" methods.

Examples of "self-organization" of this kind are numerous.

The Order of Life

To label them is easy, although gratuitous. To understand them is a long way off. Exclusively reductionist tactics will never get us there, if they persist in "going it alone"; nor, on the other hand, will sheer verbal soporifics. What the task calls for is, first of all, a job of thorough conceptual overhauling and renovation. It requires that we drop self-imposed blinders and admit to view the higher perspective of the whole—not just as its bogus literary versions, but its hard scientific core, expressed in such phenomena of emergent collective order as I have illustrated. The venal preoccupation with bits of the materials of nature as such—with "what there is"—must give way to a broader concern with the manner of their operation and use—with "how it all works". And, in this shift of emphasis, one discovers that all the bits hang together; that they are all intermeshed in webs of subtle interactions forming domains or subsystems within the overall continuum of the universe.

To emphasize the "systems" character of the dynamics of living entities, I have, ever since 1923, couched their description in terms of the concept of "fields". Lest this symbolic term again arouse alarm, let me decontaminate it instantly by the following simple example (Figure 4). Let us take a circumscribed body, depending for its maintenance on active exchange with its environment; for instance, an egg in a pond, a cell in a tissue, a human individual in society. Then let the unit multiply into a few more units; they all continue to have a share in the common interface of exchange and communication with the medium. But let the number of units keep on increasing, whether by subdivision or accretion,

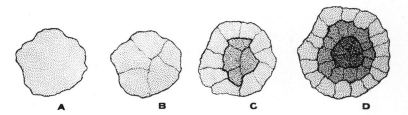

Figure 4. Diagram of progressive diversification of initially equivalent units in the course of multiplication.

and all of a sudden a critical stage arises at which some of the units find themselves abruptly crowded inward, cut off completely from direct contact with their former vital environment by an outer layer of their fellows. The latter thereby acquire positions not only geometrically intermediary, but functionally mediatory, between the ambient medium and the now inner units. From then on, "inner" and "outer" units are no longer alike. A monotonic group of equals has become dichotomized into unequal sets. With the emergence of the distinction between innerness and outerness, the 1+1 = 2 rule becomes inapplicable.

The train of events to follow such a "differentiation" of a radially symmetrical core-crust dichotomy is easy to envisage. Interactions between the "outer" members and their newly established "inner" neighbors would expose to another set of new conditions any fresh units arising subsequently in the intermediate zone between them, and hence call forth in them a third type of reaction. Moreover, polarized influences from outside would impose an axiate pattern upon the group. Thus would ensue a train of sequelae of ever-mounting, self-ordering complexity. In all these steps, the fate of a given unit would be determined by its response to the specific conditions prevailing at the site in which it has come to lie, those conditions varying locally as functions of the total configuration of the system—its "field pattern", for short. This principle—long recognized empirically as a basic criterion of systems but not always fully appreciated in its implications—is commonly referred to as "position effect".

The main point to bear in mind is that none of the component members of the group, all erstwhile alike, can know their future courses and eventual fates in advance; can know whether they would become "inner" or "outer" or "intermediate". Nor does it matter for the resulting pattern of the complex as a whole, as is best illustrated by the process of twinning. By cutting in two the cluster of cells that constitutes an early embryo or an organ rudiment, one can obtain two fully formed embryos or two fully formed organs, the way the sorcerer's apprentice, in trying to kill the water-carrying broom by splitting it down the middle, got two busy whole brooms instead. What had been destined to form a single typical organism or organ has yielded two instead, each half assuming the organization of a well-proportioned whole (see below).

An inorganic model of this process is, for instance, a sitting drop of mercury. Its convex, lens-shaped form results from equilibrium between opposing sets of forces—gravitation and adhesion, tending to spread the mass, and cohesion and surface tension, tending to hold it together. Disturb the equilibrium by cutting the liquid drop in two, and each half immediately restores its own equilibrium by assuming a convex lens shape. But freeze the original lens-shaped drop solid before cutting, and then bisect, and each half will retain its former shape of half an oblate; the dynamics that do the remolding in the liquid drop are still at work, but deprived of their free mobility, the elements can no longer yield.

The example of twinning is just ony illustration among many for the thesis that strict determinacy (or invariance) of a collective end state is fully reconcilable with indeterminacy (or variance) in detail of the component courses or events. My early inculcation of the "field" concept into biology has aimed at no more than at offering a semantic therapeutic against the spread of that epidemic of myopia and constriction of the visual field, which has left so many burning problems in the life sciences unattended. The "field" is a symbolic term for the unitary dynamics underlying ordered behavior of a collective, denoting properties lost in the process of its physical or purely intellectual dismemberment. Being descriptive of a property of natural systems, it must not be perverted into a supernatural principle; the study of those properties is, of course, an empirical task and not a literary pastime.

The field concept is apt to lead to a rational picture of progressive orderly diversification within and in reference to a given pattern of initial boundary conditions ("fields"), essentially in dichotomous steps.[8] It readily explains why we must look to the pattern as a whole as "determining" the behavior of the parts, rather than the other way around, for it is only by their position within the total configuration of the collective that the parts become marked "inner" or "outer", with all the further specifying consequences this entails; no member of the group knows its place until it has learned it. This is what endows a systemic pattern with that high degree of stability and invariance, regardless of the absolute number of component subunits and of the detailed manner in which they are initially distributed or even artificially reshuffled.

It can readily be seen that bisecting the model of Figure 4, (e.g., in the multicellular stage), in placing into outer positions cells that would have come to lie innermost, redirects radically the fate of all the cells left in each half, while the proportionate overall pattern (the "inner-outerness") of the original individual system reappears faithfully in full integrity in both. The only function of this very elementary model has been to illustrate in principle the mode of operation of a dynamic system in which macrodetermination of a group coexists with microindeterminacy of its members. For the rest, the model is far too crude to have direct biological applicability. Its main deficiency is the monotony of the boundary condition (simple and homogeneous border of exchange), which in a living system is always superseded by ordered anisotropy and inhomogeneity, expressed at least in axiation and polarity. Each of the original surface districts of the ooplasm might, however, be looked at, for the moment, as resembling something like that model.

To cite a practical example: cutting an egg in two, which can be carried out in many forms of animals, has actually often yielded the development of two whole, typically formed, and fully equipped animals, one from each half, resulting in a set of "identical twins"; but as one can easily understand, this happened only provided the cut had been so oriented as to apportion the mosaic map of ooplasmic districts symmetrically between the two halves, leaving each with a proportionate representation of the original field pattern. An asymmetric cut, by contrast, dooms the half that has been shortchanged in the deal to remaining an abortive freak, even though we know that every one of its component cells contained the full complement of genes and would have been quite competent to cooperate in the building of a typical embryo, if it had only had the necessary local "cue" that would have told it what particular response would be the ns in its particular location. For instance, if an amphibian egg half has failed to receive part of the ooplasmic surface sector containing the "cueing" field for the cells that are to form the spine, associated musculature, and kidney of the embryo, the cells developing in it will naturally fail to produce those structures, even though they might have done so if they had happened to be in the other half. Moreover, since the precursor tissue of those axial organs is an essential accessory for the elaboration of overlying ectodermal tissue into a nervous system, the freakish halves will be devoid of nerves, too. They missed their cues.

Yet, what are "cues" if not another form of "information"? Therefore, the current fashion of entrusting the genes with a monopoly on "information" necessary for the building of an embryo is bound to find itself caught short. For evidently, besides its full complement of "genetic information", each cell needs still additional "topical information" derived from the field structure of the collective mass. How otherwise could any unit know just what scrap from its full grab bag of inside information to put to work at its particular station in order to conform to the total harmonious program design? Clearly, left solely to their own devices, the individual cells and their entrapped genomes would be as incapable of producing a harmonious pattern of development as a piano with a full keyboard would be of rendering a tune without a player, or if mechanical, without a stencil with the punched-in score. To sum it up, in whatever phraseology one may choose to couch it, the basic postulate of a dualism of interaction between coarse-grain field patterns and fine grain gene responses is solidly founded on experimental and logical grounds.

Before going on, having referred to field action as "system dynamics", a brief identification of the specific meaning of that term, used throughout this book, seems called for (for a fuller discussion, see reference (C)).

SYSTEMS: THEORETICALLY FOUNDED

On the theoretical side, there is a strictly logical test for the identification of a system. It rests on the nature of the interrelations between the units conceived of atomistically through primary abstraction as isolated, separate, and autonomous. As pointed out earlier, when we reverse out steps from analysis to synthesis, we can identify unequivocal correlations between the behaviors of two such units (x, y) once we have recognized them as mutually dependent. If we then test a third unit (z), whose properties we know, in its dependence on either of the two others, we might arrive at an even higher synthetic insight, explaining x+y+z, etc., by stepwise additions. Yet note that this would apply only for those particular cases in which our original primary abstraction has been empirically validated; that is, on the premise that the abstracted entities have actually been proved to be relatively autonomous. The fundamental distinction of a system is that this premise definitely does not apply as far as the relations among its constituents are concerned. Let us assume, for instance, a triplet of units, x, y, and z, each of which depends for its very existence upon interactions with, or contributions from, the other two. Then, obviously, we could not achieve a stepwise assembly of this triplet the way we did before by first joining x to y and then adding z; for in the absence of z, neither x nor y could have been formed, existed, or survived. In short, without the *integral* coexistence and cooperation of all of them, that whole constellation would not occur except fortuitously and non-recurrently. This theorem reminds one of the "many-body problem" in physics (see later).

Regardless of the pertinence of this comparison, however, the fact remains that, in empirical study, processes in living systems reveal themselves as just such cooperative "many--body"-networks of persistently recurring interacting group performances. It is impossible to elaborate here this summary statement any further, but a few simple examples might help to clarify its meaning. Systems of this type of "physical wholeness" can be simulated by inorganic analogies. A self-supporting arch is one example. Such an arch can never be built by simply piling loose stones sequentially one upon another, for at the curvature they would start to slip off. To stabilize an arch as a self-supporting structure, a keystone must be present. In other words, an arch can only exist in its

The Order of Life

entirety or not at all. Statically, it is a system. Of course, human imagination has found ways of building arches, piece by piece, by cementing every brick firmly to its neighbor mechanically or by building first a scaffolding to hold the row of bricks in place in their unstable array until the keystone has been inserted to join the two halves and give the total structure its static equilibrium. But those devices are contrivances of the resourceful human brain, itself a living system, which thereby achieves a feat that could never have been accomplished without such constructive intervention and help from another system. In other words, system begets system. "Omnis organisatio ex organisatione" (see later).

This conclusion leads right over to a more proximate example in living systems; namely, the reproduction of the macromolecules in the living cell. Even though this process is commonly referred to as "synthesis", it is radically different from what goes under the same name in inorganic chemistry. If chlorine and hydrogen are brought together, they will combine to hydrochloric acid, even if none of the end product is there to serve as model. By contrast, the assembly of simple constituents into complex macromolecules in organic systems always requires the presence of a ready-made model of the product or, at any rate, a template of the same high degree of specificity, to guide the proper order of assemblage. The best studied case is, of course, the transcription of genes, segments of a string of deoxyribonucleic acid (DNA), into a corresponding sequence of ribonucleic acid (RNA), the orderly array of which is then translated into a corresponding serial pattern of amino acids in the formation of a protein.

Although this copying process of patterns and its various derivative manifestations, such as the highly specific catalysis of further macromolecular species through the enzymatic action of proteins, is often referred to by verbs with the anthropomorphic prefix "self", these processes are no more "self"-engendered than an arch can be "self"-building; for in order to occur at all, they require the specific cooperation of their own terminal products—the enzyme systems which, being indispensable prerequisites for all the links in the metabolic chains, including those for their own formation, thus close the circle of interdependent component processes to a coherent integrated system. Only the integral totality of such a system could with some justification be called "self-contained", "self-perpetuating", and "self-sustaining".

THE LIVING CELL: A SYSTEM

These brief theoretical contemplations lead over directly to the practical consideration of a living cell. It is impossible to convey a reasonably accurate conception of a living cell by static illustrations on a printed page or in museum models, even when supplemented by verbal description. As a matter of fact, the frozen immobility and immutability of those textbook illustrations have led to such abstruse misconceptions of a living cell that any portrayal would have to dwell more on what the actual cell is not, than on what it is. Unquestionably it has been that lack of realistic first-hand acquaintance with living cells which has been the source of some of the rather fictional current ideas, models and speculations regarding "the" cell, which, while innocent, have not always been innocuous in their effects on theory formation in the life sciences. Therefore, on many occasions, I have shown motion pictures of living cells in action under a variety of controlled experimental conditions. The purpose was to free the common mental stereotype of a cell from the strait jacket to which its static textbook illustrations have committed it. It is true that there are many specialized cell types in which the cell body encases itself in a rather rigid envelope, much as a caterpillar in a cocoon. Bacteria, plant cells, red blood cells, bone cells, are some examples. However, cell life, stripped down to its essentials, is best studied either before the cell has encased itself or after it has been liberated from its captivity in a densely packed tissue. This is done by the method of explantation or tissue culture, in which cells can be observed, manipulated, tested, and experimentally explored in their most basic behavior and reactivity. In studies of this kind, the vitality and viability of the cells under study must be meticulously preserved. Since this puts a natural limit on the scope of such investigations, one complements the limited knowledge attainable from live cells by killing, fixing, sectioning, and staining the cell to render it amenable to microscopic inspection, and beyond the limits of the resolving power of the microscope, about two orders of magnitude further down, to examination under the electron microscope.

These morphological methods have opened to our view a microcosm of microscopic and submicroscopic structures in various arrays of great regularity, specific for each kind and state of cell. But he who is not constantly shuttling back and forth between the observation of the living cell and the pictures of its dead, preserved inventory, is apt to forget that the latter gives us only momentary transient pictures of a system in incessant change. Thus, what we recognize as static form must be regarded as but an index of antecedent formative and transformative processes, comparable to a single still frame taken out of a motion picture film. Of course, taken all by itself, a static picture fails to reveal whether it portrays a momentary state of an on-going process or a permanent terminal condition. Unless this ambiguity is fully appreciated and constantly borne in mind, one runs the risk of mistaking the static picture of the cell for evidence of a mosaic of well-consolidated structures.

A cell works like a big industry, which manufactures different products at different sites, ships them around to assembly plants, where they are combined into half-finished or finished products, to be eventually, with or without storage in intermediate facilities, either used up in the household of that particular cell or else extruded for export to other cells or as waste disposal. Modern research in molecular and cellular biology has succeeded in assigning to the various structures seen in micrographs specific functional tasks in this intricate, but integrated, industrial operation. There is a major flaw, however, in the analogy between a cell and a man-made factory. While in the latter, both building and machinery are permanent fixtures, established once and for all, many of the corresponding subunits in the system of the cell are of ephemeral existence in the sense that they are continuously or periodically disassembled and rebuilt, yet always each according to its kind and standard pattern. In contrast to a machine, the cell interior is heaving and churning all the time; the positions of granules or other details in the pictures, therefore, denote just momentary way stations, and the different shapes of sacs or tubules signify only the degree of their filling at the moment. The only thing that remains predictable amidst the erratic stirring of the molecular population of the cytoplasm and its substructures is the overall pattern of dynamics which keeps the component activities in definable bounds of orderly restraints. These bounds again are not to be viewed as mechanically fixed structures, but as "boundary conditions" set by the dynamics of the system as a whole. I am deliberately phrasing the conclusion in this symbolic and rather vague form in order to leave it noncommittal for future more precise specifications, as the symbolic language of systems thinking matures.

SYSTEMS: OPERATIONALLY DEFINED

Pragmatically defined, a system is a circumscribed complex of relatively bounded phenomena, which, within those bounds, retains a relatively stationary pattern of structure in space or of sequential configuration in time in spite of a high degree of variability in the details of distribution and interrelations among its constituent units of lower order. Not only does the system maintain its configuration and integral operation in an essentially constant environment, but it responds to alterations of the environment by an adaptive redirection of its componental processes in such a manner as to counter the external change in the direction of optimum preservation of its systemic integrity.

A simple formula which I have used to symbolize the system character of a cell in the living state[9] could be applied as well to systems in general. It sets a system in relation to the sum of its components by an inequality as follows: Let us focus on any particular fractional part, A, of a complex suspected of having systemic properties, and measure all possible excursions and other fluctuations about the mean in the physical and chemical parameters of that fraction over a given period of time. Let us designate the cumulative record of those deviations as the *variance*, v_a, of part A. Let us furthermore carry out the same procedure for as many parts of the system as we can identify, and establish their variances $v_b, v_c, v_d, \ldots v_n$. Let us similarly measure as many measurable features of the total complex, S, as we can, and determine their variance, V_S. Then the complex is a system if the variance of the features of the whole collective is significantly less than the sum of variances of its constituents; or, written in a formula:

$$V_S \ll (v_a + v_b + v_c + \ldots \ldots v_n)$$

In short, the basic characteristic of a system is its essential *invariance* beyond the much more variant flux and fluctuations of its elements or constituents. By implication this signifies that the elements, although by no means as single-tracked as in a mechanical device, are subject to restraints of their degrees of freedom so as to yield a resultant in the direction of maintaining the *optimum stability of the collective*. The terms of "coordination", "control", and the like, are merely tautological labels for this principle.

To sum up, a major aspect of a system is that while the state and pattern of the whole can be unequivocally defined as known, the detailed states and pathways of the components not only are so erratic as to defy definition, but, even if a Laplacean spirit could trace them, would prove to be so unique and nonrecurrent that they would be devoid of scientific interest. This is exactly the opposite of a machine, in which the pattern of the product is simply the terminal end of a chain of rigorously predefined sequential operations of parts. In a system, the structure of the whole coordinates the play of the parts; in the machine, the operation of the parts determines the outcome. Of course, even the machine owes the coordinated functional arrangement of its parts, in last analysis, to a systems operation—that of the brain of its designer.

Let us return now to our main line of argument which culminated in the postulation of a dualistic principle of mutual interaction between the dynamics of field patterns, on the one hand, and appropriately matching selective responses of the genome, on the other.

ORGANIC FORM: THE FIELD-GENE DUALISM

Unquestionably, that postulate must prove distasteful and disturbing to those who, craving final and quick answers to nature's problems, have found the surest satisfaction of that urge in paring and trimming broad questions down until they fit into the narrow frame of answers available as of the moment. To stay within that frame as long as is compatible with facts and logic is a time-honored and sound scientific practice, particularly as the frame is always taken to be somewhat elastic. But when, stretched to the limit, it still is found to be too narrow to hold its content, we must give in to nature and build a wider frame. In the broad view of experience with the development of organisms, the frame set by the dogma of gene monopolism has proved to be a strait jacket. The evidence is cogent. It is of two types.

There is first the evidence against the logical possibility of deriving an orderly space pattern, such as that of an organism, from merely summing up algebraically the activities of discrete and freely mobile units, whether these be genes or larger subcellular entities carrying transcriptions of the genetic code and their derivative translations to proteins; for how could scalar small-grain units add up all by themselves to the large-grain overall geometric (in dynamic view, vectorial) order of the product of their joint construction? The question is, of course, rhetorical. It is not unlike asking for the probability of sand grains, free from constraining guidance, arraying themselves perchance in a straight line—and not just once coincidentally, but systematically and consistently. This should be borne in mind in assessing current efforts to salvage the atomistic "information-carrier" concept of development by the recent interesting demonstration of the presence in egg cytoplasm, hence outside the nucleus, of "information--carrying" DNA. It is immaterial whether such macromolecular relatives of the chromosomal genes are nuclear emissaries or autonomous "plasmagenes", for which latter there is substantial proof from several sources. All that is relevant to our basic argument is that, being afloat in dispersion throughout the cytoplasmic space, they are no better fit than their nuclear counterparts to provide the scaffolding for the coarse-grain spatial ordering of the fine-grain developmental processes, which they help carry out.

The second line of evidence is experimental, based on the fact that the cell population of an embryo in sufficiently early stages of development can in many instances be reshuffled quite haphazard without any trace of the disarray showing up in the collective product, which still turns out to be a typically organized individual.[10] This proves, first, that the individual cells at the time when the population was scrambled were equivalent and indefinite as to their destinations, but above all, the fact that the group dynamics which subsequently defines their destinies has, as an entity, the necessary definition and constancy of pattern to guide them. The term, "regulation", commonly given to this phenomenon, merely raises a problem; it contains no answer. In fact, if the question is whether the development of an embryo can actually be put together from piecemeal contributions of its genetically informed cells, then the simple expedient of just calling their contributions "regulated", and stopping right there, begs the question. For it is arguing in circles to pose that it is feasible to compound that which "regulates" from the sum of that which is being "regulated"—the macroregularity of a formative field pattern, from the microproperties of spatially disarrayed work units.

POLYMORPHISM: SAME GENES—DIFFERENT FORMS

For further evidence, let me turn back once more to the split egg yielding a pair of twins called "identical". Stemming from the same zygote, their cells naturally contain identical genomes. But this is about as far as the "identity" usually goes. The point is that in their overall morphology, "identical" twins can really be the most dissimilar specimens of a given species; for they are essentially mirror images of each other, either entirely or partially, in that the normal asymmetry of organ structure and location is expressed only in one, but is reversed in the other. The name of *situs inversus* (site inversion), by which the phenomenon is known, is insufficient as it refers only to location, while in reality the inversion affects the inner structure, too, (e.g., the direction of coiling in spiral organs or organisms). Thus, the two members of such a set of twins are as incongruous as a left and right hand. However, apart from this major morphological discrepancy, the detailed features bear rather close resemblance, which is a telling test and index for the instrumental role of the genes in the specific cellular reactions of which those features are the visible signals. By the same token, this makes, of course, the nongenic matrix of the basic morphogenetic pattern stand out all the more clearly in contradistinction.

Space structures in mirror-image configurations are well known from crystallography. Designated as "enantiomorphic", they represent alternative equilibrium states of systems which have different properties along their different axes in space. The two states need not occur with equal frequencies, for any bias imposed by environmental conditions (e.g., a gradient) can favor one state over its counterpart. Instances of similar structural enantiomorphism have been observed in development. They have furnished one of the strongest incentives for postulating a conceptual parallel between the morphogenesis of organisms and of crystals. This is not the place to ponder over whether the comparison will turn out to be analogy or homology, but the sheer fact of a formal similarity is indisputable. As it is pertinent in the present context, it calls for a few further comments.

Ross Harrison, in a series of classical experiments, showed that the early limb bud, a cluster of cells protruding from the embryonic body wall, is primed by prior interactions to form a limb, and nothing but a limb, although none of its component cells "knows" as yet just what role it will be called upon to play in that performance; whether to contribute to the building of a given piece of skeleton or a particular muscle or tendon, or plain connective tissue. The bud is in possession of a "limb field". A limb bud transplanted to any place on the body at that stage sprouts a limb. Moreover, in line with our previous conclusions on the twinning of early embryos, such a limb bud, when split in two, proceeds to give rise to two whole limbs.[11] Yet, as with "identical twins", such twinned limbs tend again to be mirror images of each other. More unexpectedly, even a whole limb bud at a stage when it is already fully and restrictively endowed with the capacity to form a limb is still ambivalent as to its pattern of asymmetry; that is, upon being transplanted, it may give rise to either a right or a left limb, the decision hinging essentially on its orientation relative to the main axes of the body, or more correctly, of the axiate pattern of its immediate surroundings.

Since corresponding results have been obtained consistently in experiments with other organs (heart, ear, feathers, etc.),[12] it is quite evident that we are faced here with a fundamental principle of "macrocrystallinity" in morphogenesis, formally akin to crystallinity in heteroaxial inorganic systems.[13] The organic morphogenetic system has initially two equivalent and mutually symmetrical (enantiomorphic) dynamic configurations open to it, but once one of the two has become established, it precludes the manifestation of its opposite. Which one of the original alternatives is to assume dominance and become expressed, is decided, in an anisotropic environment with a consistent asymmetry of its own, by the statistical bias imposed by the latter, and in an essentially isotropic environment, by chance. But the outcome is always distinctly one or the other alternative, and never anything in between; there are no intergrades. Would any thoughtful person claim that this type of overall morphological regularity and lawful predictability could be derived from nothing but the summated separate "information" bits of thousands of freely shifting individual cells contained in a limb bud, heart tube, or feather germ, devoid individually of any geometric regularity to which the geometric asymmetry of their collective product could be referred?

One of the reasons for laboring this point is the confusion that has arisen from the ambiguous statements in the literature that "asymmetry of shape is genetically determined", based on the demonstration of Mendelian ratios in the offspring from matings between left-coiled and right-coiled snails. Actually, the recorded ratios pertain not to any primary determination of asymmetric shape in the observed individuals, but rather to the general asymmetry in the anatomy of their mothers. It is not the genic constitution of the egg itself, but the asymmetry in the maternal environment of the egg, which biasses the egg to effectuate one of its two enantiomorphic alternatives of morphogenesis to the exclusion of the other; much in the same manner as the polarity of the body wall surrounding a transplanted limb bud tips the scales between capacities for either right- or left-handed configuration without "determining", i.e., designing, limb configuration as such. Let us not confound the designing of a pattern with the mere favoring of an available one over its competitor. After all, one can choose between a right and a left glove without knowing the first thing about how to make a glove. From all those tests and other similarly pertinent ones, the thesis of the primacy of developmental design over the know-how of the fragmented genetic craftsmen, in interaction with which it is carried into effect, has emerged not only inviolate, but strengthened.

We might go on to reinforce it further by pointing to alternatives of dimorphism of even greater disparity than those between mirror images; for instance, the phenomenon of "homoeosis". In some forms of regenerative repair of a mutilated organ, the cell population of the stump engaged in the regrowth can give rise to either of two distinctly different, not intergrading, structures. The remnant of the stalk of an amputated crayfish eye, for instance, can regenerate either a new eye or an antenna. The severed antenna of a praying mantis can sprout, instead of the removed antennal end, a leg; and so forth. The substitute structure is usually one that is typical of a neighboring region. The decision hinges in each case on the level of amputation, that is, again on a topical relation between the formative cell group and its body surroundings: the more of the original organ has been left behind in the stump, the greater is the probability that the regenerate will restore the original form, whereas the closer the cut is made to the base of the structure, the higher is the incidence of the homoeotic formation. Evidently, the rules of how to build an eye, as well as those for the building of an antenna (or in the second example, for an antenna and a leg), but nothing else, are inherent in the patch of growth-stimulated cells at the lesion; yet which of the two alternatives is to prevail depends on where the cell group stands within the larger complex.

Just what it is that tips the scales in favor of one course over the other is unknown. Reference has been made on occasion to critical levels of hypothetical "gradients of metabolic activity" of the area or of concentration of "morphogenetic substances", whatever that may mean, on the presumption that each of the two competitive processes would have its peculiar environmental optimum, which would lie below the critical level of activity or concentration for the one, and above that level for the other. Mindful of the fact that regenerates sprouting close to the base of an appendage tend to develop into structures germane to neighboring regions one could also speculate that the decisive factor is the relative local strength of the overlapping fields of neighboring appendages; the field strength of the genuine local structure (eye or antenna, respectively) diminishes in proportion to the extensiveness of the amputation, thereby enhancing the opportunity for the "intruding" field of the adjacent appendage to assume dominance and to trigger in the growing cell mass the latent pattern of its own type.

Examples of such dynamic competition between adjoining fields are familiar to embryologists but the inclusion of homoeosis in the same category is strictly conjectural. The crux of the matter remains that a given mass of proliferating cells at a relatively advanced stage in life, being by no means single-tracked toward a single preset destination, can still be switched into one or the other of two radically different morphogenetic courses, each of typical organ pattern, the decision between them depending on their dynamic relations to the larger system in which they are enmeshed.

Significantly, the same sort of "homoeotic" substitution of one type of organ by one of a neighboring type, shown here to occur during regeneration has also been observed in primary development as a result of genetic mutation; for instance, the mutated fruitfly *aristopedia* regularly develops a leg in the place of the antenna. The fact that the very same substitution of a leg for an antenna can come about as the result of both a mutational aberration of the genome and of a rather unspecific variation of epigenetic conditions (i.e., the level of amputation) points to a similarly unspecific common denominator of both phenomena. One might argue as follows. In regeneration, the leg-versus-antenna decision might be attributable to the passing of a critical threshold in the ratio of metabolic conditions, which normally is balanced in favor of antennal over leg differentiation, except toward the base of a developed antenna, where it becomes reversed. In order to bring the genetic case into line, one could assume then that the mutation in question entails a general shift in the metabolic balance of the whole animal in the direction normally prevailing only at the antennal base of the developed animal; the leg process, ordinarily subdued in the head region, would thus be given dominance over its antennal competitor even in earlier developmental stages.

For lack of factual information, there is no telling how close this explanation comes to the truth. No matter how close or remote, however, it does make the fundamental point that is at issue here. It shows that although the mutational genetic transmutation of an antenna into a limb is a spectacular and impressive event, the underlying genic alteration need have absolutely nothing to do with the *patterning* of "limbness" or "antennaness" as such. By analogy, your freedom to choose on the market between an apple and a pear tells you nothing about what has made them different in the first place.

In conclusion: if one considers that the dynamic structure of a field pattern is a continuum, which tends to retain its integrity as it is passed on ad infinitum to the offshoots of the growing protoplasmic continuum, which stretches from primordial living organisms through present ones into the future in undisrupted continuity, and if one chooses to designate such direct transfer as "inheritance", one comes to

The Order of Life

realize that "genetic inheritance" covers only part of the story of the maintenance of continuity between successive forms of life, between generations of organisms, and between generations of cells within each organism.

CELL POPULATIONS: MODELS OF FREEDOM

It ought to have become clear in the course of my discussion that arguments about "determinism" versus "indeterminacy" without specifying attributes, as if they were categorical and absolute opposites, are senseless. Determinacy is subject to a scale of degrees, which cannot be applied except in specific reference to a particular level on the scale of hierarchical orders. If one moves down that scale from top to bottom by actual observation and experimentation, instead of the usual extrapolating from the bottom up by postulation and exhortation, one cannot miss the "nonreductionist" lesson that macrodeterminacy (in the sense of predictability) can demonstrably exist on a higher level without being based on any correlated microdeterminacy on the next lower level.

The significance of this conclusion for a theory of development, and, in a wider perspective, of life processes in general, can readily be grasped. Its full acceptance, however, will still have to contend with the addictive relapse into the purely formal and unrealistic notions of development engendered by stereotyped book versions that are so schematized that mental pictures abstracted from them bear scant resemblance to the real thing. It might be worthwhile, therefore, as a sample of a corrective, to add here a few plain facts, selected at random, which might help to loosen up the prevailing all-too-rigid image of development. [14]

One of the facts to bear firmly in mind is the tremendous rate of cell proliferation during growth, coupled with such great variability in the distribution of the descendant cells that any sample of tissue or organ one might take from a given individual is absolutely unprecedented and unique, that is, unlike any other sample in the detailed distribution of its cells. Biopsy samples from exactly corresponding sites of two identical twins reveal no "identity" in their cell configurations. The casual variability in cell production is enormous. For instance, the two "equivalent" daughter cells of a given mother cell can differ in dimensions and content by as much as one to three, quite haphazard. In many tissues, proliferation becomes gradually restricted to cells in given layers or areas, another test of their deriving secondarily some specifying cues from their positions in the total geometry of the cell mass (e.g., "at the surface"; or "innermost"), while the other cells, crowded out from those privileged stations, switch into tasks of labor or of surplus reserve; but no cell in such a tissue "knows" up to the moment of positioning which of the alternatives will be its lot. The destiny of no individual member of a given cell population can be regarded as prefixed in predictable detail till it has gotten near the end of its course of life.

In some organs, cell multiplication ceases completely after a certain period, for instance, in the case of nerve cells. In other organs, such as gut and skin, proliferation continues throughout life. Digestive cells in the lining of our gut divide once every few days and then are shed into the lumen; the cells of our incessantly growing skin flake off as horny scales. The number of successive cell generations in the life span of a single individual thus can go into the tens of thousands—as many as all the generations of man in recorded history. Yet, skin and gut retain essentially their standard pattern of structure and function unaltered despite the continual changeover of the populations of cells composing them. Parts of the blood cell populations turn over even faster; a recent estimate lists the daily birth and death rates of certain white cells as of the order of billions.

While on the subject of cell death, it should be pointed out that from the very start of embryonic development, the occurrence of degeneration and death of cells in growing organs is not only common but quite extensive, net gain of mass resulting merely from the excess of cellular birth rate over death rate during the earlier phases of the individual life span. In later phases, when cell birth rate declines, the balance becomes less favorable, and from a certain point on, a

The Order of Life

tissue may lose more cells than it can replace—a fact which, among others, contributes to the physical symptoms of aging. Surprisingly, this critical point is reached earliest in the nervous system, for in all higher vertebrates, including man, multiplication of nerve cells terminates shortly after birth (by which time the human brain alone has acquired a nerve cell population from ten to one hundred billion). From then on we have to make do with that initial endowment of brain cells for the rest of our lives. Not only that, but nerve cells keep dying right throughout life, our native stock of brain cells said to dwindle by more than a hundred thousand cells each year. Consider, for a moment, the implications of this fact. Cells that are parts of the neuronal network that is instrumental in all our neural and mental functions, drop out haphazard, almost one every minute, without this loss at all disrupting or affecting the integrity of our patterns of perception, coordination, behavior, and memory, or our sense of identity. Compare this dispensability with the fatal effect of removing the smallest critical cog or coupling from a clock-like machine, and you will realize the absurdity of ascribing to a single cell a critical determinative role in the orderly operation of a living system—in the present instance, the nervous system; in the earlier one, the formative dynamics of development.

    I hope these few examples from the book of facts, not fantasy, will have helped to put the central issue of this book into sharper relief by illustrating in yet another context the fundamental characteristic of a living system, which is: the property of the system as a whole to build, operate, and maintain itself in a state of desirable orderliness, predictable from its repetitive and systematic recurrence and endurance, in spite of the continued flux and infinitely greater range of variation, hence, unpredictability, of the behavior of its constituent units, which carry out the building, operation, and maintenance. The actual courses taken by the individual units are so unique and nonrecurrent in detail, from moment to moment and from case to case, that taken separately, they bear no sign of relevance to the *orderliness* of their collective product. The order of the system rests on the *coordination* of all their courses. The individual courses, therefore, are not "determined" by what the individual units know from the start, but by coordinating signals from the total complex, guiding and holding them to the overall design. We need not go further here into the various special coordinating subsystems the multicellular body develops secondarily to keep the integrated operation of that growing cell mass manageable, such as the nervous system, the hormone system, the homoeostatic maintenance of the composition of the body fluids; for in principle, each one of these subsystems operates within its own scope by the same rule of integrative dominance that the higher system exercises over its components.

FREE SYSTEMS: PROBLEM SOLVERS

This principle of *conservation of overall pattern by the coordination of the erratic flux of the component elements*, already plain to ordinary observation of the steady state, is even more conspicuously documented by the reaction of such systems to outside disturbances of such novelty that they could not possibly have been experienced before by any cell, brain, or organism, and hence could not have been provided for in the genetic program of the species in anticipation of such emergencies. Practically all instances of healing and repair of damage, of compensation for deficiencies, of adaptations to forcible changes of the inner or outer environment, and the like, bear signs of improvised *ad hoc* adjustments.[15] One could include here the experiments, mentioned earlier, proving the development of half an egg into a whole animal. However, since the fission of eggs does on occasions occur in nature, it could be, and has been, contended that to cope with natural accidents of this kind, evolution has endowed eggs of exposed species with a program for a "reserve machine" to be set in action in an accidentally separated egg half. When it was shown, however, that one could also do the opposite, namely, fuse two whole eggs and end up with a single giant individual, "machine preformism" had to concede defeat, for nature could not conceivably have provided eggs with spare equipment for the nonsensical act of forming only half a body.

The Order of Life

In general, however, coordinated pattern-conserving responses of living systems to artificial experimental interferences merely accentuate the evidence for the general principle of systems dynamics spelled out above, but otherwise need not be regarded in a separate class than the simple deductions to be drawn from ordinary variance. Whether or not one wishes to compare this basic property of a living system with Heisenberg's indeterminacy (or "indeterminability") principle, laid down for the subatomic realm, is a moot question of philosophical predilection. It is really beside the point as long as the very existence of supra-atomic and supramolecular levels of determinacy of a state of order that cannot be derived exclusively from statements about the lower levels is properly recognized in its legitimacy. Under whatever name, the acceptance of the principle profoundly affects the way of thinking about living systems. It does so in a liberalizing sense by liberating biological concept formation from the excessive constraints imposed by an outmoded faith in microprecise linear cause-effect sequences as "mechanisms" of natural events. Instead of letting a macroevent come about by a rote concatenation of innumerable bundles of microevents, in which each microevent is considered to be linked rigidly and exclusively to its linear predecessor and successor—head-to-tail, as it were—we must concede to the microevents far wider degrees of freedom. But we must also view them as crosslinked and interwoven all along in a web of dynamic interrelations of such nature that any deviation of any given microevent from the standard pattern of the total system will appear coupled with a resultant re-equilibrating deviation of opposite sign of the ensemble of all the other microevents. Clearly, if microevents had consciousness, the realization that they have, at any moment, not just one, but many degrees of freedom open to them, would give them the experience of "choice," supplanting fated predetermination by acts of decision with a total design in view.

Obviously, the simile of "conscious knowledge" by every part of a system as to its own place, state, and movement, and, what is more, of places, states, and movements of all other parts, is no more than a picturesque analogy; it does not portend a relapse into panpsychistic metaphysics. Of course, the mere mention of this behavioral correspondence will make most life scientists squirm. In fact, I suspect that fear lest such parabolic analogies develop into spurious homologies, pretending to explain the real thing, has been at the root of a certain distrust of biologists in the "systems" character of living objects, even though the latter are after all the star witnesses for the "systems" concept. Thus, to allay that fear instantly, let me point out, by way of example, that any system to which we ascribe elasticity behaves precisely as outlined above. Distort the most complicated rubber figure whichever way you want, and it will, upon release, resume its original configuration; just as if each particle of the mass had "known" precisely the direction and magnitude of its particular dislocation relative to all others and from that information had taken its bearings for the return trip to its erstwhile location.

To be sure, this is a very primitive example, especially as the envisaged system is a monotonic one; that is, one in which all relevant variables are of a single category—mechanical strains and stresses. But the model may help in overcoming the emotional resistance to systems-thinking as such, as if the latter necessitated abdication to ultrascientific or supernatural control agencies. Once adopted for monotonic systems, there would seem no further barrier in principle to letting it expand to encompass multivariant systems, such as the living systems, in which unitary effects result from the combined action of multiple and manifold forms of energy, interacting conjointly in coordinated fashion, "as if" in conscious goal-perceiving cooperation.

In recent years, the behavior of systems has been brilliantly reproduced—not to say, mimicked—by the performance of computers, displaying "self-preservation," "self-regulation," "learning," and many other features of behavior resembling those of the intelligent system that has built and programmed those machines—man. Enthusiasts have gone so far as to equate brains with computers. This is not the place to discuss the shortcomings of that equation, except to state that even if it were valid, it could not serve as model for biological systems altogether, not even for the individual brain cell; for

The Order of Life

although the electric wiring scheme of electronic computers may seem near enough to resembling the network of fiber connections among nerve cells transmitting messages in electrochemical code, one must remember that the cell itself executes all its integrative and unitary systems functions without possessing anything resembling circuitry and wiring diagrams. The objection that such devices might be there in some occult state still undetected is countered incontrovertibly by innumerable experiments, in which cells have survived and kept on operating without impairment despite the most violent stirring and disruptions of their interior.

There do exist also in many organisms machinelike mechanisms which have evolved as specialized automatic safety devices and which operate by built-in feedback controls in the manner of a man-made thermostat (e.g., elementary stereotyped reflex responses). The main point to bear in mind, however, is that those are very special cases and by no means prototypes of the far more fundamental and inclusive type of systems dynamics of organisms, which can maintain integrity and pattern amidst internal flux and external disturbances without such preprogrammed anticipatory machinery. To elevate the simple stereotyped reflex (e.g., the knee jerk on tapping the tendon below the knee cap) to the rank of a universal model for the operation of the nervous system (e.g., creative thought or even the singing of a song) is just as unwarranted as if one were to proclaim the red blood corpuscle of a mammal, which has no nucleus, cannot grow or reproduce and is doomed to die within weeks, as the paragon of a cell. The free systems mode of operation of man's creative intelligence can contrive contraptions that mimic human behavior—mechanical dolls or puppets manipulated by humans—but these creations are not thereby elevated, by arguing in reverse, to true models of systems behavior.

CONSOLIDATION: SYSTEMS BECOMING MECHANISMS

In their substance, all these comments reiterate that freely operating systems have an evolutionary tendency toward progressive consolidation. Just as streaming water, having freely grooved its bed, becomes constrained into its self-created channel, so those parts of a systemic network which are engaged in the systematic repetition of a given standard performance, running a standard course, tend to become fixed and structured into stereotyped tracks; thus they acquire aspects of automatic chain reaction mechanisms at the price of losing degrees of inner freedom and adaptiveness. Pictorially, this trend could be compared to the freezing of a liquid (cf. the mercury model above). In the liquid system, the equilibrium condition of minimum surface makes the elements of the liquid body shift freely until the body has attained the requisite shape—e.g., that of a sphere. When such a drop is cut in two, each of the disequilibrized halves will take on sphere-shape, reminiscent of a whole twin forming from a halved egg. But let the drop first freeze and then be cut, and you will see each fragment retain the hemispheric shape it had before, because the solid crystal bonds among the elements have deprived the system of those degrees of inner freedom that would permit it to resume the equilibrium configuration of a sphere. Systems dynamics has been superseded by rigid structure.

In a similar sense, the "self-differentiations" of isolated parts of embryos are to be viewed as secondary derivations of prior, more "fluid", states within wider dynamic systems; habits, as frozen residues of erstwhile freer problem-solving dynamics of the mind; and institutions, regulations, and mechanisms of control, crystallizing in the system of a free society, evidently as congelations of established habits. The primary, self-preserving order of a system thus seems to drift relentlessly toward a state of dynamic disintegration into separate fractional compartments, eventually highly structured, mechanized, autonomous, and correspondingly inflexible and unresponsive to the rules of order of the original unit system. If held together in the old pattern by inertial forces and cements, such an assemblage of fragments may still offer a sham semblance of inner unity, but any test of adaptive responsiveness will show that its erstwhile free dynamic unity has vanished.

We need not concern ourselves here with those special

The Order of Life

modes of tertiary *reintegration* which, in certain instances, supervene and make up for the secondary disintegration of systemic self-sustaining order and which are exemplified by the emergence of the nervous and hormone systems concomitantly with the progress of "self-differentiation" in the embryo. The crucial point to remember is that the actual demonstration of certain machinelike, rigidly concatenated devices and performances in organisms must not be misinterpreted to mean that an organic system is nothing but a bundle of such mechanisms—in other words, is no "system," in the strict sense, at all. A biological mechanism must rather be regarded either as a fortuitous simulation of a systemic feature, like mimicry, discovered by coincidence and stabilized by evolution, or else as a specialized secondary derivative of a primary and genuine system activity. This change being unidirectional, from the wider range of degrees of freedom of the system to their progressive and eventually complete repression in the emancipated "mechanized" parts, it would be unreasonable to contend that the primary systems property of life processes could be reassembled conceptually by sheer summation of analytically describable, minutely determinate, single-tracked "cause-effect" sequences. Yet, this is essentially the mental image conjured up by unqualified references to given characters, expressions, or actions of living systems as "genetically determined."

SURVIVAL: BALANCE BETWEEN RIGOR AND PLASTICITY

Thus, while the turn my comments have been taking might seem to have veered from the chief issue, we find ourselves now back right at the heart of it. So, let us summarize. What I have tried to document is the need for subjecting the concept of "determinacy" to a critical dissection in the realistic light of what we actually know about life processes, particularly development and nervous functions, before we can speak of "determination," genetic or otherwise, with any cogency. It turned out that in that undifferentiated generality, the term is rather meaningless; it is useless for scientific purposes and devious in its further generalization to those extrascientific areas of thought and practice which rely on scientific underpinnings. What we came to discern on closer inspection was that determinacy is not an all-or-none proposition, but must be specified as to type, degree, and "grain size."

Schematically, we can distinguish three types.[16] (1) A consistently ordered pattern of a complex system may be determined as a whole, without the behavior of its constituent units being determinate, that is, preprogrammed each for its individual task: "macrodeterminacy *without* microprecision," for short. (2) There are, however, as we noted, also group performances in which the courses of all components are rigidly predetermined in such a manner as to yield automatically a collective product of predesigned orderliness; this would connote "macrodeterminacy *through* microdeterminacy," characterizing mechanisms in contradistinction to systems. (3) Intermediate between those two types are complex systems, the components of which are neither so uninstructed as in (1), nor so firmly structured and autonomous as in (2), yet are still subject to cooperative interactions guided by the systemic pattern as a whole. Metaphorically, type (2) could be compared to someone being given definite road instructions on how to get to a given destination, whereas (1) refers to one who asks those who know the land for directions all along the route; when encountering an unmarked detour, the former would evidently be completely lost, while the latter, not having known the difference between a standard route and a detour in the first place, would make inquiries just as he would at any fork, hence stay on course. Such road forks symbolize the multiple choices open to the living system to meet the unpredictably varying details of each course.

From this we can predict that although no one has yet come up with a systematic listing of biological processes under those three categories, any such attempt would end with only a single category, namely, (3); for survival hinges on a proper balance between the efficiency of rigid design, as under (2), and flexibility for adaptive change, as under (1). In concrete terms if we designate systems dynamics by the letter "S," and mechanisms by "M," a behavioral act would, for

The Order of Life

instance, appear as follows. Starting from the top down, an instinct or habit is composed of a rather rigid sequence (M) of motor performances, each of which is in itself unified and internally variable (S), but also employs rather rigid reflexes (M) in its execution; all of this being done through the agency of cells, which, as we pointed out earlier, operate as systems (S), although in part again through rigid mechanisms (M), as in the transcription of the genetic code from chromosome to messenger units and consecutive translation from messenger to protein. One could cite many more examples for this mixing of degrees of determinacy in biological processes. Even so, there is no question that the list would be much larger if past research had not selectively concentrated on the M-bands of the total spectrum of biological processes to the exclusion of the much tougher, though fundamental, problems of the S-continuum, which have been largely bypassed.

The realization that no relevant operation of an organism consists entirely of mechanisms—M-chains—has the most profound bearing on our concepts of a living system. A chain that is interrupted and open-ended is functionally no chain at all. Accordingly, whenever a microprecisely determined reaction chain merges into the statistical pool of free systems dynamics, that signifies not only the end of the chain, but also the end of its determinative role: a supposed "minidictator" is then reduced to the role of member and participant in a patterned macrodetermined operation, in which its separate track gets lost; far from "determining" any piece of the planned pattern, it can only contribute means to its execution. If one chooses to compare the degree of precision of fine-grain microdeterminate processes (e.g., DNA→RNA→Protein) with the degree of resolution attainable by an optical instrument, one recognizes at once that any M, on entering into systems dynamics, incurs an instantaneous loss of definition and resolving power—in terms of information theory, degeneracy. This is the same as saying that even if microprecisely programmed chain reactions were set off in separate linear tracks as "determiners" for organized end products, their aim would get blurred on the way in the trackless stretches of the network of systemic anastomoses.

To go back once more to the specific example of development, the course from gene to character is not in category (2), the uninterrupted single-tracked reeling off of a schedule of preprogrammed events, as which it was once conceived in the early formalistic phase of genetics. The simplicity of that primitive concept could not endure in the face of mounting practical experience to the contrary. It was observed that single genes affect a whole set of diverse somatic characters, and that conversely the effects of many separate genes would converge on a single common target character. Through several such transitional phases, a "one gene—one character" concept gradually evolved into its exact opposite, asserting that in reality each character is potentially influenced by every gene and each gene, conversely, is potentially of influence on every character. This in no way disputes the fact that often the change of one single gene may find expression in the alteration of a single conspicuous assay character; for any localized input into even a trackless system (for instance, a local distortion of a network) can yield a response signal so unequivocally corresponding to the input that it looks as if input and terminal manifestation had actually been connected by a direct traffic line. Yet the multiplicity and profusion of demonstrated relations between genes and characters have ruled out definitively the putative singularity and linearity of gene tracks in favor of a network concept.

What this change of view, however, has failed to take into account, is the progressive blurring of definition that occurs in such a network from two sources, both consequences of the vagaries of epigenetic events. One source is the unpredictability of the precise course any given cell among the many hundred billions of an average mammal will take in its development; the individual cell thus can have no prescience of what local circumstances it will meet and have to respond to from moment to moment. The other source, related to the former, lies in the fact that the layers diagrammed in Figure 1 (p.11), through which all communication to and from the gene must filter, are likewise subject to capricious developmental fluctuations. This compound variance reduces the degree of detailed predictability (i.e., determinacy) of the component processes of development to the point where it would become inconceivable that an orderly product could ever emerge from the resultant blurred muddle, unless there were systemic integrative dynamics in operation to keep the total variance down to the narrow limits expressed in the invariance formula for cells (p. 41). Even if such systemic dynamics had not been demonstrated, they would have to be postulated.

The dualism of, on the one hand, systemic field continua, representative of *vectorial* order, and, on the other hand, the discrete *scalar* packages of genes, with which the former are in interplay, does satisfy that postulate. However, ascribing the stability of overall pattern to hypothetical "regulatory" or "coordinating" genes does not. For regardless of the extraordinarily regular array of the genes themselves, the cells that carry them roll around without fixed bearings. The patterned form of their ensemble would thus have to arise *de novo* from a mere algebraic shuffle of numbers of gene vehicles—geometry to be derived from algebra—which obviously would presume the intervention of magic—a proposition which a science worth its name would rather shun. To recount the evidence for this dualism, scattered throughout this essay, seems unnecessary, but perhaps a glance at the presumable origin of the prevailing thesis that genes have a monopoly on the dynamics of development might be of help.

WHAT GENES DETERMINE: SUBSTANTIVES OR ATTRIBUTES?

I sense that the thesis of a genetic monopoly is rooted in a semantic ambiguity, namely, the emphasis on characters (plural) in contradistinction to character (singular), the former referring to separate criteria or marks of an entity (e.g., green; pointed; sluggish; forked; etc.); the latter, to the whole ensemble of properties of an entity (e.g., personality; the quality of form; adaptability; etc.). One can readily understand how this has led to the illusion that character is nothing more than a bunch of characters. An individual thus had to appear to genetic analysis not as a subject, but as a conglomerate of attributes. Thus one identified, for example, genes bearing on black *vs.* blond hair and called them, for short, genes for "black" or "blond" (but then, what is hair?); or similarly, in the eye, genes for "blue" *vs.* "brown" (what is an eye?); for harelip *vs.* unsplit lip (what is a lip?); for color vision *vs.* color blindness (what is vision?); and so forth. What one had actually established was a correlation between gene differences on the one hand and *differences* between entities on the other. Yet, shorthand usage gradually abstracted the differential attributes from their substrata, keeping the characters "black," "blue," and "split" in view, while forgetting about their carriers; i.e., hair, eye, and lip, perhaps in the expectation that they will likewise in the end prove decomposable into a collection of attributes; but attributes of what?

Let it be noted that this question is rarely being posed, and if so, has never been answered with the same scientific rigor and logic that has distinguished the truly admirable advances of the genetics of differentials. Of course, one hears the popular vernacular use phrases like genes for "eye" or for "health" or for "intelligence." But no one in the know used to take this kind of language seriously, and the matter of the common core of the phenomena whose differential features were definitely gene-related was allowed to lapse into obscurity. In fact, it has suffered further obscuration in our day by being treated with the magic cure-all of "information" verbiage, which in this case proves to be just a placebo as a pretentious cloak for "lack of information." This criticism reflects in no way on genuine "information theory," which must not be blamed when its name is being taken in vain. What happened was that, as genes had been rigorously demonstrated to account for modifications of attributes of eyes, such as eye color, eye size, eye function, and so forth, genes were now suddenly also postulated, or rather proclaimed, to manage the very process of eye formation. How? By virtue of containing, as the story goes, the "information" for eye formation. What does this mean in realistic terms? Absolutely nothing. The problem of "eyeness," as such, like that of the essential orderliness of any other organic form, remains what it has always been—an

The Order of Life

honest question mark. It seems gratuitous to submit that the mere substitution for it of the word "information" is actually an answer, and thereby to vindicate the sole and exclusive authority of the gene in the "determination" of living patterns.

In this connection, it is important to consider that the very methodology of genetics makes it virtually impossible ever to test such a claim of omnipotence. All direct genetic tests are based on comparing *differences*, either between parental strains or between a standard stock and a mutated strain. Testing beyond the range of crossbreeding fertility and mutability is therefore impossible, which confines the direct proof of gene involvement to distinctions within races and species, while leaving the major features that are shared by larger groups of organisms untestable. Indirect evidence, extrapolated from immunological determinations, is more inclusive in that it has revealed parallels between the phylogenetic kinship of different forms of animals, on the one hand, and the degrees of similarity among their gene-dependent antigenic proteins, on the other. But since even these tests pertain only to criteria known to be gene-dependent, anyhow, and evidently cannot be extended to the nondifferential common core of formative dynamics—to the overall character of forms of life rather than their differential characteristics—the thesis of a basic macrodeterminate, patterned, field dynamics, of which the discrete indeterminate scalar units (cells or subcellular complexes) with their complements of genes are the executive tools, cannot be disputed on genetic grounds.

It is impossible, of course, in this brief excursion into the land of genetic fundamentals, to counter all the arguments to be anticipated in the defense of the prevalent creed. For example, one might be confronted with an impressive array of "genetically determined" deformities as witnesses testifying to an exclusively genic foundation of morphogenesis. The retort would have to be that all the known examples in that category have been, or can be, attributed to either quantitative deficiencies or imbalances in the genic repertory through which patterned formative dynamics are translated into overt formed products.[17] In other words, the genic aberration does not create a new pattern, but merely modifies the expression of an existing one. A limb may turn out longer or shorter; a whole part may be missing because the gene-dependent chemical kinetics of the cellular labor force was too slow to furnish a necessary supplement at the critical time; a growing limb bud may have become forked so as to yield extra digits (hyperdactylism), because, of all things, gene-dependent fluid production of the lining of the brain (hydrocephalus) had become excessive, spilling from the ventricles into the tissues and coming to rest as a blocking blister at the growing tip of the limb bud. Developmental deviations, resulting from disharmonies in chemical kinetics, similar to those based on defective or ill-matched gene combinations, can also be produced by external disturbances of the epigenetic course of development; for instance, by drugs (see the pathetic cases of babies with defective extremities from thalidomide-treated mothers) or by virus infection (e.g., measles in the mother during pregnancy). Remember also the "phenocopies" in animals mentioned earlier. Yet, not a single instance has come to our notice in which the ordered ground plan for *form* as such, which all those variable formed products have in common, could have been documented to be likewise the work of "gene activity."

In conclusion, "genes" are not everything in life, even though nothing in life seems in the long run capable of proceeding without them. Water, for instance, is also indispensable for life, yet we no longer subscribe to the doctrine of Thales Miletus that water is the sole building block of the universe. The whole misconception, as one now readily perceives, is intimately tied up with the naively mechanistic "single cause-single effect" picture of nature, which I dealt with in the early portion of this essay. In a facetious mood, a vitamin was once defined as "a substance that makes you sick if you don't eat it". More seriously, but with equal pertinence, a "cause" (or gene) is something *without* which some "effect" (or character) which you expect *fails* to occur, while something else occurs instead. To turn the sum of such negative statements around and fashion

The Order of Life

from them a positive doctrine of plenipotency (of causes or genes) seems to me a reprehensible somersault of logic.

Therefore, instead of dwelling further on the origin of misconstrued concepts of genetic determinism and genetic omnipotency, I shall concentrate now rather on their consequences. Some of them are distinctly favorable in that they further the dramatic progress of genetics proper, since in any subject, limitation of the field of vision is conducive to mastery. Some others, however, ought to give us pause. They are the catchwords which are apt to lure the life sciences gradually into blind alleys. Although such deviations are mostly passing self-correcting episodes of purely scientific concern, the ones at issue here are broader and more ominous because of their moulding effect on the foundations of human thinking and action—or inaction. By way of example, let me briefly survey four of the areas of thought thus affected: the concept of the attribute "alive"; the problem of the so-called "origin of life"; the antithetic distinction of "genetic vs. environmental factors" (the nature-nurture problem); and the question of the scientific validity of "freedom of choice" and "self-determination".

LIFE: MATTER OR PROCESS?

In scientific terms (as distinguished from broader philosophical, religious, or private definitions), the question of "what is life?" seems to resolve itself into a listing of the irreducible minimum number of objective criteria necessary to characterize a "living being". But once one settles down to try it, this task turns out to be fraught with ambiguity. In the first place, objectively we do not know of such a thing as "life" at all; all that our experience presents us with are living organisms, which we compare with dead organisms (as well as with permanently "nonliving" things), vesting the difference by abstraction with the term "life". One of the most striking aspects of that difference is that in the living organism, in contrast to the static lifeless one, things seem to happen all the time—exchange of stuffs and energy with the environment, movements, growth, reproduction, etc. This is, indeed, correct, but only for the "living" organism, not necessarily for any "live" one. To choose an example, there are, quite aside from seeds and spores, certain animals among the lowly types of spiders (tardigrades) and worms (nematodes), which can be kept in a state of complete dehydration, in strong ultraviolet light, at boiling-point temperature or in liquid nitrogen (-195° C) without showing any signs of life for years, but which, when brought back eventually into "livable" conditions, recover from their dormancy and resume the regular activities of the living. Accordingly, there is a state of being alive, yet not living; but would we want to call it "nonliving"? Plainly, the animal in complete stasis could not be told apart from a "dead" one; only the test for "revival" can tell. So, what is life?

Let us next look at cut flowers. Bereft of roots, their life span is reduced, which we express by calling their residual existence "survival", rather than life. But what of the case of further comminution of a plant or even animal down to its constituent cells, which then, if transferred to appropriate nutrient media, can go on living, growing, multiplying for practically indefinite periods. Such "tissue cultures" have been called "immortal". But looking closer, one recognizes that the component cells, which do the "living", have very limited life spans; in fact, whenever a cell divides into two daughter cells, the mother cell ceases to exist as an individual: is dead. What actually is perpetuated interminably in such a colony is not at all the life of the individual cell, but the continuity of generations, the existence of the population, which, of course, possesses none of the biological criteria of life except in a metaphoric sense.

I have cited these rather bewildering examples of contradictions as proof of the futility of asking whether or not a given item or phenomenon should be dignified by the attribute of being "live" without qualifications. In this light, the question of whether "life" can be created is gratuitous. One can only ask, at best, whether "live" organisms can be generated from strictly "nonlive" elements without any intervention of preexisting "live" organisms or of their organized products. This rigorous phrasing immediately disposes of the

The Order of Life

popular question whether viruses are alive. No virus has ever been shown to be able to "do" anything on its own; the cooperation of a living cell is indispensable. The nature of this dependency is again misrepresented by the customary phrase that viruses "reproduce" only inside a livng cell; in reality, they do not reproduce at all: they are being reproduced by the cell through its own machinery, although the configuration of the virus serves as template or model for the shape in which the cell will replicate it. In principle, though in a much simpler version, the shaping by an immunologically competent cell of an antibody that matches the macromolecular pattern of a foreign antigenic protein is in a similar category, yet one would not ascribe vitality to the protein molecule.

This kind of reasoning extends of course to all species of molecules, including DNA, which in one of the most epical achievements of modern biology has been revealed to be the physical embodiment of that array of units previously referred to symbolically as "genes". No denigration of the superb accomplishments of its detailed analysis is implied in deprecating its misinterpretation and abuse, of which one finds unmistakable signs in semiscientific and popular discourses, but also in some serious scientific speculations, on the "origin and potential re-creation of life"—the "synthesis of living cells", as it is called. The successful synthesis of DNA from nucleotides, its small molecular building blocks, has actually been publicized as heralding the first major step toward the "synthesis of life" in a test tube. But so was also the announcement of the first synthesis of an organic molecule, urea, by Wöhler more than a century ago. What is the innocent layman and the broad public to make of such claims?

First, a reservation: is it at all conceivable that life has come about, and could once again be made to come about, in steps? The answer to this question will be considered further below. Second, a restriction: the synthesis of DNA could be effected only with the specific aid of a highly organized product of a living cell—an enzyme—whose own production and configuration depended reciprocally on the coding action of an existing DNA; which is a vicious circle. Third, a comment: even granting, against reason, the possibility of artificial DNA synthesis from scratch, the suggestion that this would constitute the inauguration of life is predicated on the thesis of the omnipotence of the genome, the very antithesis of the postulate of systemic order, as set forth in the preceding discussion. The difficulty of granting to primordial DNA dispensation from dependence upon the ordered protoplasmic system, which its more recent offshoot is known to need as source of its indispensable operational prerequisites (such as, for instance, the essential enzymes, energy sources, etc.), was neatly met by conferring upon DNA by a semantic sleight the power of "self"-replication, something unknown in the vocabulary of preorganic chemistry.

A fourth and final argument, however, strikes even deeper. I have already indicated that to arrive at a sharp positive definition of life is an elusive task. Nevertheless, one can at least narrow it by establishing what life definitely is *not*. One can, for instance, submit that life is not matter; this is clearly reflected in the conventional—and incidentally, rather objectionable—distinction made between dead and living matter; in other words, the same "matter" in different states. If life is not matter or substance, then evidently the "elements of life", which scholars have tried for ages to pinpoint, cannot be elementary particles of matter under whatever names—including "molecules". If we want to describe life in its most general and universal aspect, we must call it a *process*—a *change* of matter and energy content in time. Therefore, in plain logic, the "element of life" can only be conceived of as an *elementary process*. To proclaim static units or compounds as "elements of life", is about as illogical as if one tried to measure velocities in grams. Moreover, the prime characteristic of that process is the systemic *order* that rules its dynamics, in contrast to random processes, much as a melody contrasts with noise.

THE "ORIGIN OF LIFE": FACTS AND FANCIES

Thus, sober and comprehensive evaluation dethrones the DNA molecule, both the hypothetical primordial one and its

The Order of Life

test tube facsimile having the dimensions of matter, from its imputed role of autarchic ruler of "life", which has the additional dimension of time. Since, as I have tried to stress, static mass cannot be converted into ordered dynamic process unaided, we must call on the pretender to abdicate. The reference here to the *order* of the dynamics of the living system should make it plain that we are not speaking of the relativistic conversion of mass to energy, nor even of the conversion of stored potential energy into free kinetic energy. Nor would it do to relegate the feat of turning a molecule into a process, as if "instilling" life into dead matter, to the various and sundry animistic, vitalistic, magic and mythological agencies that have been conjured up for the purpose; and to confer spontaneity (as implied in the prefix "self-") and other anthropomorphic powers upon a static molecule is not a bit more scientific. *Deus ex machina* expedients are not admissible in science. Is there a scientific way left open, then, to explain the origin of life without recourse to such parascientific expedients? Despite some ingenious speculation brought to bear on this question in recent years, no answer in scientific terms is in view that might pass the test of rigorous scientific critique. Conjectural solutions thus far presented, although brilliantly imaginative and appealing, have failed that test for reasons briefly outlined in the following.

The proposition is simple: after all, life on earth is a fact, and since our planet was formed in a state decidedly unfit for the survival of any hypothetical primordial germs of living beings that might have been carried over from other stellar bodies, we must conclude that either such germinal contamination has come over later, after the earth had become livable, or else living organisms have actually arisen totally *de novo* here on earth. In view of the hardiness of primitive organisms (spores, etc.; see the earlier reference to live, but nonliving, states), passage through the extreme climate of outer space cannot be wholly ruled out, but there is no trace of evidence that this may actually have happened, either. Moreover, even if it had, the problem of the "origin of life" would thereby merely be shifted from our globe to the universe, and, hence, remain scientifically unresolved. Viewed realistically, therefore, "life" (i.e., "the irreducible minimum of criteria of a living being"; see above) seems to have emerged right here and science cannot dodge the task of exploring and explaining how. Great minds have struggled with the task, yet have come up with no more than partial answers. Since, as I have stressed superabundantly, no living organism or living process can be conceived of as a piecemeal assembly product of independent fragments or events, fragmentary solutions hold no solution at all, for they miss the crux of the problem of "life", which lies in the systemic order of the component processes. This argument of the nonfractionability of life militates with equal force against the notion of its stepwise origin. Whatever the minimum existential criteria are that must be conceded to the simplest living organisms, they constitute an indivisible complex of interrelated properties—indivisible in the sense that lacking any one of them, the rest no longer satisfy the minimum specifications of a living system.

It will have become clear by now that all attempts at solving the problem of the origin of life are inextricably trapped in the ambiguities of the various connotations of the term "life" itself, to which I have already briefly alluded. It is not surprising that, in the resulting haze, one often resorted to stripping the object, "life", of as many of its integral properties as was necessary to bring it within view and grasp of a solution. The fact that this mental trimming act deprived the living system of features indispensable for life, thereby killing (speaking symbolically) the very object of the search, has largely gone unnoticed. To re-erect the object in its full magnitude, a concrete inventory of the essential and indispensable attributes of an elementary living system is necessary. Without going into as much detail as the subject would merit, I shall attempt to give a sample list. Like all our speculations about primordial life, it is, of course, based on extrapolations from the elementary properties of living systems as we know them, reduced to essentials. In carrying out this backward projection, it is well to keep in mind a sharp discontinuity in our mental operations when, first, we retrace evolution from the present to the most primitive ancestral living form and then go further back from there to

The Order of Life

reconstruct the prehistory of the latter. For the evolutionary transformation phase, we apply reasoning in continuity, deriving each later sample stage methodically, by evidence or analogy, from the preceding one. For the prior phase of the emergence of the first living forms, we have no similar guides for rational extrapolation. To apply the term "evolution" to both of them is a misnomer for the following reasons.

As in genetic theory (see above), the focus of attention in evolutionary theory has been again on *differences* of form and function, while the unchanged stock of basic common properties has failed to receive commensurate emphasis. Yet, as the studies of comparative physiology, cytology, and biochemistry have revealed, the conservative endowment *in common* to all living organisms, from the highest to the lowest, is so impressively large that the French term *"transformisme"* seems to express the principle underlying phylogeny more accurately than does the conventional term "evolution". In essence, evolution could be characterized far more appropriately as a vast series of "variations of a common theme" (to be sure, rising to ever greater exaltation of emergent novelty of combinations) than as a sequence of basically novel themes. If the most primitive (that is, least specialized) single-celled organisms known are taken as the closest contemporary representatives of primordial living forms, their equipment for the performance of the basic vital processes (e.g., marcromolecular synthesis, metabolism, cell multiplication, contractility, excitation, membrane functions, respiration, secretion, pigmentation, fiber formation, and others) is so remarkably congruous with that of the highest (that is, most artfully specialized) animals and plants that the whole course of evolution begins to look like a continuous reshuffling and more propitious recombination of an extremely limited number of immutable performance elements, contained within an equally restricted canon of systemic rules of order circumscribing their cooperative interactions. In metaphorical terms, one is reminded of the structure of language, which from an extremely limited assortment of letters forms a larger and more varied, but still narrowly circumscribed, vocabulary, the unit words of which, through the flexible device of syntax, are freely recombined to sentences that make sense—sense, which if novel, might be called "idea". The essence of creation lies in the imaginative recombination of old elements to constellations of striking novelty of order, rather than in the addition of new basic elements—in man's mind as well as in evolution.

In surveying phylogeny in this perspective, one is impressed as profoundly by the perseverance in absolute constancy throughout the ages of the basic vocabulary of life as one is by its dramatic permutation for the creation of emergent novelty in evolution. The countless varieties of proteins in living beings are all composed of no more than a score of amino acids, just as all printed words are composed from only twenty-odd letters of the alphabet at the typesetter's disposal. It does not matter, in the present context, that in terms of the concept of the "genetic code", each amino acid is itself the "translation" of a word spelled out in sequences of nucleotide letters, the derivative words then being combined to standard phrases called protein. Note that this leaves wide open the whole question of syntax: the proper choice and ordering of phrases to form sentences and language that make sense.

Even a protein, however, is as such a dead and static unit which can "do" nothing by itself except disintegrate and liberate energy as toll to the ever growing pool of entropy. The captive energy thus freed had previously been invested in it in the act of building its ordered, i.e., highly improbable, structure. But invested by whom? Well, in the present, by coupled processes in which solar energy, stored in food as fuel, is transferred stepwise to the building site by the catalytic device of enzyme systems. Yet, since all enzymes themselves are proteins, we are confronted with a simple *quid pro quo*, as the formation of fresh protein presupposes the cooperation of preexisting protein. This logical hurdle to deriving a primordial protein from a protein-free oceanic soup is not lowered by making protein itself a secondary dependent upon nucleic acid primers, because the priming action of the latter is, in turn, dependent upon the intervention of enzymes, hence, protein.

The Order of Life

Ingenious conjectures about the cosmic climate on and around the infant earth have proposed direct energy supply from powerful radiations as possible substitutes for the otherwise inescapable vicious protein-to-protein circle. I am not competent to judge the merits and prospects of those challenging speculations about the emergence of highly complex organic compounds from mixtures of simple inorganic molecules—in fact, the feat has been reproduced artificially on a modest scale (amino acid synthesis). However, I question whether the matter is relevant to the key problem of the origin of living organisms; for, as I explained above, a molecule, regardless of how complex it may be, is dead, the difference between a live and a dead body being not one of their molecular inventories. The primordial "element of life" must be visualized not just as a blob of organic *matter*, but as a system of interdependent *processes* resulting from orderly interactions in organic matter. To establish the minimum number of such indispensable interactions is an empirical task and not to be left to fancy. For guidance, one might proceed as follows: One would first list all those essential properties and processes that are shared by living organisms on all levels of complexity, from microbes to man. One would then examine the degrees in which they are dependent upon one another and scratch from the list those which practical experience with growth, propagation, and maintenance has shown to be omissible without fatal consequences for the rest. The common denominator left as residue of this stripping operation might then be accepted as the irreducible minimum set of cooperative ingredients whose co-existence would be indispensable for even the most primitive living process.

PRIMORDIAL LIFE: TEST OF PRIMORDIALISM

In stressing the indispensable coexistence of those ultimate prerequisites of primordial life, I anticipate the inescapable conclusion that life, as we know it, cannot possibly have developed gradually in consecutive steps, but must have come about in a single major cataclysm. But before arguing the point, let me sample a few more of the basic properties of living systems that have persisted through the ages of organic evolution without essential change and which, therefore, illustrate the basic stock-in-trade of all life, including that of its earliest exponents. If protein emergence, just outlined, were really the single solitary crucial step to life, from which "all further blessings flowed", one might perhaps acquiesce in a wait-and-see attitude. But it is not. To be sure, the enzymes involved in the fundamental operations of capturing, transferring, and storing energy (aerobic and anaerobic metabolism) are the same throughout, and they are proteins; so are those engaged in the functions of digestion, excretion, contractility, etc.

However, this common inventory of macromolecular species represents only one aspect of the nature of living systems. The crucial fact is not just the presence of this or that interactive or synthetic event, but the coordinated linkage of all of them into a network of integrated effects. Selective exchange of substances with the environment requires a bounding interface or membrane: all organisms have it. The lipids and carbohydrates forming it are much the same in all of them. The blood pigments for oxygen transfer in animals are structurally similar and they, in turn, resemble the chlorophyll of plants. The mechanisms for mechanical work (e.g., pumping or locomotion), for impulse conduction, for cell division, for the production of fibers and fibrous fabrics: all operate according to much the same principles in all forms. Does this not clearly document the obstinacy with which a repertoire of a few basic themes has been preserved in its integrity throughout the course of organic history? And would it not then stand to reason to take this repertoire to be that minimum *sine qua non* for life, primordial as well as recent, which can be compellingly deduced from our empirical knowledge? Yet, the acceptance of sheer plurality of primary essentials in coexistence is only part of the postulate. The salient feature is their enmeshment in a network of *interdependence*. They are not a package of independent items, as unrelated to each other as the content of a soldier's emergency kit, but each makes indispensable contributions to

The Order of Life

the existence and operation of the others, just as its own existence and operation, in turn, depend inseparably upon theirs.

To expand this fundamental point in detail would only mean to hark back to the main "systems" argument highlighted in this whole book. Suffice it, therefore, to cite one example, the pertinence of which is often overlooked. Biochemists, by and large, study organic chemical interactions (e.g., enzyme reactions) in physically orderless or deliberately disordered states (test tube solutions or tissue pulp). This is a technical necessity and a lot can be learned through these methods. But they fall far short of reproducing the way things happen in the living system: they can show, at best, what might happen, but not what actually does happen in the organized state; for in an organism, chemical reactions occur, in the first place, not ubiquitously and indiscriminately, as in a mixture, but rather in critical patterns of localization and segregation, and, second, without the pampering and nursing care of a learned experimenter. As for the first point, localization and segregation presuppose some dynamic compartmentalization of processes—sorting out as against intermixing—which one might subsume under the general designation of "structure". Here we encounter another one of those inseparable circular entities denoting systemic properties, to wit: organized patterns of chemical activity, predicated on structural organization, but the latter, in turn, established and maintained by chemical activities of synthesis and energy supply.

This comment leads right on to the second point. In order to be able to study any separate component of the chemical works of a cell or tissue in isolation, an investigator must astutely provide it not only with just the optimum environmental conditions (temperature, electric states, etc.) but also all the necessary material ingredients which normally are furnished by the internal chemistry of the cell itself and, incidentally, many of which the investigator must borrow from a cell in prefabricated form to have success. Of course, this is a laboratory procedure which has in special circumstances been anticipated by evolution in the phenomena of mutualism, symbiosis, and parasitism; but one must never lose sight of the fact that it always implies barter or theft between two living systems, a loss by one being repaired by another. One cannot reverse the case to use it as rationale for the presumption that the minimum essential endowment of the first true organism could actually have been built up stepwise in consecutive fractional acquisitions, when there were no other organisms around to be robbed of the missing supplements.

No glossing over the two cardinal points discussed in the preceding paragraphs can hide the insurmountable hurdle they present to any theory of a stepwise origin of life. The hurdle consists of the axiom of interdependence among the plural component processes of a living system. The magnitude of the plurality is wholly irrelevant. Our sketchy list above of indispensable components may be too restricted or too generous. There may have been no more than three. Let us examine, for instance, such a triplet of mutually linked processes, a, b, c, each one providing some essential prerequisite for the existence and operation of the two others. It is evident, then, that such a triplet could never have come into existence stepwise as a + b + c, for no two of the members of the set, (a + b) or (b + c) or (c + a) could have, by definition, existed unless the third member (c) or (a) or (b), respectively, were already present, each with its indispensable share. The problem is analogous to the "many-body" problem in physics, as pointed out earlier. Just as the latter, it reemphasizes the frailty of obsolete micromechanistic cause-effect chain reaction models, epitomized by the a + b + c + . . . sequence, whose inadequacy to describe not only the operations of a living organism, but the very origin of life on earth, calls, by the same token, for the recognition of "systems" as valid primary entities of the universe—disorder and randomness becoming secondary mental abstractions arrived at by disregarding orderliness; but such a broad leap into the realm of philosophy would take me on ground where I would feel unsafe and ill-equipped to keep from sinking. Nevertheless I shall make a confession at the end.

Yet, the conclusion about the "origin of life" on earth I

The Order of Life

cannot shirk. With all due respect for the valiant and imaginative efforts made in the past to reconstruct how primordial living beings might have "evolved" by progressive additions to an erstwhile cluster of inorganic matter, advancing from nonliving through something like gradations of one-eighth living, one-quarter living, three-eighths living, and so on, to full-fledged vitality, I cannot bring myself to adopt that concept in the face of my experience to the contrary, gained in half a century of intimate investigative and interpretative study of living organisms, the main conclusions of which have been embodied in this treatise. As for the "origin of life", my conclusion must remain that, as the phrase correctly indicates, it has *"originated"* rather than *"evolved"*, that is to say, it must have arisen or emerged, perhaps in many places, in a single sweeping step upwards on the hierarchical scale of order in the universe; a step assembling several erstwhile independent nonliving components, potentially matched for mutual harmonious cooperation, into a new higher integrated entity capable of enduring as such.

The fortuitous occurrence of such a singular event, in which all pieces fall into their appropriate places in subordination to the new overall order of their collective system, appears to be so infinitesimally small that one either has banned it from scientific concern by calling it, extrascientifically, a "miracle", or else in scientific frustration has negated the proposition. Need science really abdicate yet? Perhaps we have indeed reached the limits beyond which pure reason cannot penetrate. Perhaps, however, our scientific view of the universe has been too narrow, including a too restricted concept of probability (see my earlier remark about the questionable primacy of randomness). To prognosticate how the future will decide would certainly mean trading scientific discipline for poetic license. It might be worthwhile, however, to speculate that modern thermodynamics, in explaining the stability of compound systems in terms of their "minimum free energy content", is opening new prospects for occurrences in which lower-order units in a common environment would combine and "fall into places" in new collective groupings of higher order, because *the free energy content of the emergent complex is lower than the sum of the free energies of the separate components* for which the joint environment is severally *suboptimal*. At any rate, this excursion brings me again up to the limits, not of scientific competence, but of my own. So, let me retrace my steps to where I branched off from the discussion of the concept of "genetic determinacy" as such.

HUMAN FATE: FROM MECHANISM TO FATALISM?

At that point, I proposed to examine the broader implications which an unwarrantedly narrow monopolistic interpretation of the concept might have for human thought and actions in general. I first touched on its potential bearing on popular notions of "life", culminating in such questions as: "Does the genetic code of DNA molecules hold the master key to the understanding of life?" The upshot has been this: If one uses one's private notion of "life", as indicated in the introductory pages of this book, following the Foreword, such questions are no more sensible than if one were to ask for the wave length of the "vibrations of the soul" in poetry. Sticking to the narrower, strictly scientific, i.e., universally communicable, language, the specific query above has been answered in the negative, but even there the target problem of "life" proved to be too heavily blurred by semantic indefiniteness to lend itself to bulls-eye hits of scientific answers. What science has been able to establish is only what life is *not*; the positive assertions of what it is or might be are still a matter of diverse factional beliefs rather than scientific knowledge, and possibly are even doomed to remain so for all times. What science definitely could do is reduce the number of discrepant concepts of living systems—the discrepancies usually referable to sectorial views from specialized corners— so as to save a confused lay public, as well as the experts in other disciplines, the spectacle of factional strife among conflicting scientific doctrines. But even then, "life" will still remain an abstraction, the real things being processes in living beings; the hope to isolate a "carrier of life" is a chimera, no less elusive than would be an attempt to extract the center of gravity from a body. The problem of the "origin of life", in

The Order of Life

critical overview, did not fare better. As in the former case, scientific statements about it were reasonably safe solely in a restrictive sense, in pointing out how life could *not* have come about—for instance, not in a succession of independent acquisitions of the basic prerequisites for living—while, on the positive side, science has to veer from its safe course and let contention take the place of unattainable evidence.

A wide margin is thus left for conjecture, personal predilection, and traditional belief—a margin within which science is out-of-bounds, as it were, to act as a guide in the affairs of man. Even within that range, science still has authority to unmask demonstrable untruth, but it cannot attest truth. It can strive to keep narrowing that range, but no sober reasoning from current knowledge would encourage the expectation that that band of scientific uncertainty will ever shrink to a line and be abolished. Nevertheless, much in contemporary phraseology dealing with "genetic determinism" might be misinterpreted by the uninitiated to mean just that: that an individual's destiny is, from his conception on, grooved, guided, or misguided, toward a blind and inexorable outcome, which it is not in his powers to direct toward a goal. If this modern version of an ancient doctrine of absolute predestination, in which the genes are simply subrogated for the supernatural predesigners of old, were allowed to take hold because of science's failure to disavow or rectify some loose or misinterpreted utterances by scientists and their acolytes, man's will to strive will lose its drive. Suggestive jargon, innocuous for knowledgeable insiders, but apt to be misconstrued by the innocent, could easily, if absorbed and propagated in that corrupted sense, turn into a signal handicap to man's responsible conduct; like a road sign to an oasis that has become turned sideways so as to point into the desert.

My whole concern with this matter stems from the fact that I seem to sense a current trend in that direction in the connotation the public mind places on "genetic determination". Laying stress neither on the latitude left to a living system throughout its life span nor on the rules of order which keep the system integrated and viable despite that latitude, we have allowed a notion of rigid predetermination for life, a feeling of an inexorable destiny, to settle in the public mind, the end of which would be the embracement of a philosophy of *fatalism*, with all its social and political overtones. Such fatalism would negate man's powers to do something about his fate, and perhaps enjoin him through social mores from even trying; and all this on the mistaken supposition that rigidity of epigenetic life is scientific gospel, while science, in fact, proves the exact opposite: the flexibility of epigenetic development and function. Thus, while science, as I attempted to show, increasingly succeeds in documenting the freedom of parts within canons of order and thereby certifies, as it were, on scientific grounds, the opportunity for self-improvement of body and mind, the abridged message of science is easily perverted in the public mind to something like: "I am a slave to my genome; I cannot break its bonds; so, why strain myself?"

Undoubtedly, this uncritical attitude of resigned fatalism would receive further reinforcement from educational, social, or political philosophies which stigmatize effort, even far short of the stress limit, as a necessary evil rather than a life-sustaining and life-improving virtue, and consequently propound that the burden of maintaining balance in a population be shifted from the consummate efforts of *all* the individuals to a disproportionately large share from the genetically more highly endowed segment.

BREAKING THE SOUND BARRIER: BEYOND GENETICS—EPIGENETICS

This is not the place for me to engage in more detailed prognostications about the sociological, and even moral, hazards to emerge from an image of mechanistic rigidity imputed to "genetic determinacy". I do wish to stay clearly on the scientific side, and from that vantage point to reaffirm simply the scientific legitimacy and proficiency of man's effort to control his fate. To be sure, his genetic endowment confines his individual potential within a finite frame or repertoire which distinguishes him from every other individual. How-

The Order of Life

ever, the *epigenetic* course of life, being decidedly not an automatic reeling off of a rigid chain of microprecisely concatenated events, offers man wide opportunity and the faculty for steering his course within that frame for better or for worse. Instinctively, man has always displayed the urge for exercising those individual faculties; for instance, in the play of infants, in sports, in the creative arts and crafts. Of course, the genetic stock, which predisposes sportsmen or craftsmen for greater proficiency in one field than another has pre-set their differential talents, but do we not know from vast experience that talent has to be developed to become achievement? That aspect of development, the epigenetic one, is not genetically "determined"—just genetically "circumscribed".

It is important for the continuity of our thinking in this connection to bear in mind that "epigenetic" refers by no means to a separate time phase, as if there were two distinct periods to life, an early "genetic" machine phase, followed by a second phase, during which the machine is perfected and run in—a misconception evoked by the Greek prefix "epi-", meaning "upon". One need only return to our initial account of development and particularly the diagram of concentric shells (Figure 1) to realize that the "epigenetic" aspect encompasses all that is non-genic in development and, therefore, in an indivisible field-gene continuum, is never absent. Most broadly one could identify it with the totality of the "environmental" interactions to which the captive genome is subjected. As one can see, such an extragenic "environment" is always present.

The same consideration points up the fallacy of the conventional verbal reconciliation between the purported *rigidity* of genotype and recorded *variability* of the phenotype by letting environmental factors amend the execution of a primary genetic design. All questions taking the form of "is this or that character genetic *or* environmental?" testify to that misdirected way of thinking, which becomes especially devious when it is applied to human behavior. It is made no more objective by assigning percentage values to the respective contributions by heredity *and* environment, as if environment were a sort of contamination, which if removed would then reveal in the residual percentage the pure genetic product in its pristine unblemished glory. This is about as unscientific as if one were to express the volume of a gas, which is proportional to the ratio of temperature over pressure, by crediting a given number of cubic centimeters to temperature, and what is left, to pressure. Genes never lack environment right from the moment at which we arbitrarily choose to start our dating of the development of an individual—the fertilized egg; and since every subsequent developmental step is contingent upon the conditions created by the preceding steps of interactions between the genes and their environment, the artificiality of looking for their respective contributions, comparable to donations of money to a pool, is plainly evident.

What has confounded thinking has been that, due to inattention to the hierarchical structure of living beings, illustrated in our diagram of shells, the term "environment" has mostly been used indiscriminately without specifying the respective boundary. Sometimes it meant the natural outer environment of the individual (nutrition, meteorological and social climate, stress, etc.), sometimes the *milieu interne* of body fluids and tissue associations, sometimes the cytoplasm around the cell nucleus, whereas in reality, as far as the genes are concerned, it comprises all of those to the extent to which they are in last instance relevant for genic interactions. From the earliest stages of development on, every cell of the body constitutes environment for all the others; every cytoplasm, for the nucleus and the cell organelles; every chromosome for the gene-strings in it. Throughout development and life, at every moment and at every boundary of this system of Chinese boxes, epigenetic decisions are being made involving choices between alternative courses and directions. Were it not for the interrelatedness of those choices with a resultant—some would even call it, metaphorically, an intention—of insuring the viability of the whole system, organisms could not exist. Accordingly, the choices must, on the whole, be rather for the better than for the worse. Since these gene-environment interactions are never interrupted, the gen-

The Order of Life

eral idea that "environment" enters the scene only when the bird hatches or the baby is born should be expunged.

FITNESS FROM STRESS: MAN'S ADAPTIVE SCOPE

A few examples chosen at random might illustrate the adaptive principle of "choices for the better" during development. The details of the vascular network, particularly the system of capillaries, are largely moulded by the direction and intensity of blood flow at the innumerable points of branching and anastomosing and not according to any preset scheme. In essence, the situation is not unlike the formation of a river delta, of which one can only tell that the water will end up in the ocean, but never what precisely the distribution and configuration of the branches will be. In both cases, increase or decrease of flow rate and pressure will alter the pattern significantly. The central nerve cells in the embryonic brain, spinal cord, and ganglia, whose nerves grow out with singular capriciousness of detailed courses, will keep enlarging if their roaming ends reach and are accepted by appropriately matching terminal tissues (muscle; sense organ; other nerve cells), but those that fail of attachment and remain "unemployed", shrivel and may die. Reciprocally, embryonic muscles lacking nerve supply likewise degenerate. The basic shape of joints in the skeleton is crudely moulded by the growth patterns of the various cartilages that act as scaffoldings for the later bones—growth patterns which, in turn, vary conspicuously with the internal environment of ions, nutrients, and hormones, themselves products of different nonosseous organs—but the convex head and the concave socket of a joint will not fit snugly together unless they are mutually "ground to fit" in actual mobile contact, comparable to fitting prefabricated clothes to a prospective wearer.

All these conspicuous epigenetic improvisations demonstrable in the embryo continue right on through life. The blood vessels, in caliber, strength, valves, and branching pattern, take on the features of the most cunning design of hydrodynamic engineering. Nerve cells enlarge in proportion to the bombardment of stimulation and shrink with inactivity. Exercised muscles grow, but without use, they wither. Bones, in their inner structure, assume configurations as cleverly arranged for optimum resistance to the mechanical strains and stresses imposed on them by load carrying and movement as only the best builders of bridges, cranes, and derricks could contrive; but whereas these functional patterns are readily remodelled in adaptive response to altered stress patterns, resulting, for instance, from fractures or deformed joints, they become completely obliterated if bones are freed from stress, as in the chronically bed-ridden patient.

Such examples of "functional adaptation", pre- and postnatal, are not only illustrative of the adaptive latitude of an organism, but are singularly instructive in one regard of special pertinence to our discussion: they refer to the physics of pattern, rather than to mere chemistry of substance, even in cases where the adaptive change is measured superficially only in terms of mass, as in hypertrophy and atrophy. In illuminating this fact, they reinforce my earlier argument that for the elucidation of the problem of organic form, myopic preoccupation with the biochemistry of the gene not only can bring no answer, but inevitably misses seeing the question. It is mainly for this didactic reason that I have omitted from the list the strictly molecular adaptations, such as enzyme inductions, and especially the immunological phenomena, which do operate on the "fine-grain" level of the gene, just like, more broadly, the processes involved in cytodifferentiation.

The instances of "functional adaptation" carry a signal lesson in proving scientifically that man can do something positive for his brain and body, if he will only try. But equally important is the reverse consideration that underoccupation of his faculties entails their loss—often irretrievably. Biology thus can make a valid case against a stress-free life. If at all attainable, it would be disastrous for individuals and society alike. As I have just illustrated, bodily organs in disuse tend to lose their efficiency, eventually approaching functional incapacitation. Therefore, just as overstressing them is harmful, even short of the stress limit

The Order of Life

where human endurance might snap, so sparing them consistent effort and exercise, even of moderate and tolerable amount, is a distinct biological hazard. The evidence of high susceptibility and low disease resistance of animals reared in a protected germfree environment, supported by the statistical indications of epidemiology that resistance to infections is heightened in proportion to the incidence of repeated subclinical exposures, can be extended to all the body functions. These facts ought to be taken most seriously as a caveat from the life sciences lest unrelenting quest for effortless existence, free from challenge and incentive, such as a prosperous society might indulge in, become a lethal tranquilizing drug of that part of mankind habituated to its use.

This being, of course, a strictly biological assessment, it would not be up to me to estimate to what extent the resulting debilities of body and mind could be offset by technological prosthetics of wheelchairs, automation, and computers. My role is merely to assert that neither our bodies nor our minds are "genetically determined", programmed and pre-equipped to live, survive, and thrive, unless we practice the epigenetic faculty—indeed, the scientifically certified commandment—of striving to *bring our latent potential to optimum realization*. This calls for the exercise of will power and self-direction by judicious choices—decision processes akin to those encountered in the routing of the component processes of development through the maze of opportunities of choice among the forks of pathways available. Can we deny to the mature brain a feature which the cell system that constituted it undeniably possessed—the *power to choose*?

"FREEDOM OF CHOICE": THE BRAIN AS SYSTEM

This brings me to the fourth and last of the samples of forebodings about the effects of misconstrued "genetic determinism" on human behavior. If centuries of learned discourse among philosophers about freedom of the will have brought no resolution, it might seem preposterous for me to enter the discussion with an aside, save for the fact that the increased reliance on scientific verdicts, or rather the exegesis of such verdicts, has made the scientist coresponsible for the eventual judgment. Since the branch of science most intimately involved is psychology, which in turn rests on underpinnings in the life sciences (neurobiology, genetics, developmental biology, pharmacology, pathology, and animal behavior), a word from that side would not seem to be out of order. As ought to have become clear from my preceding discussion, that word is the decided affirmation of the *scientific legitimacy of freedom of choice*, as corollary of the indeterminacy of pathways in integral systems dynamics—valid as for all living systems, so for that special central part: the brain.

As an insert, let me give to those unfamiliar with the tally of elementary brain constituents, the sobering summary argument against any innocent notion of the brain as a micromechanical precision machine. Like all organs, the brain consists of a population of cells derived from the single-celled egg by a vast network of processes—proliferation, specializations, movements, differentiations, interactions, and cross linkages of startling complexity and diversity—all of it running its developmental course from egg to baby to adult with the degree of incidental individual variance among the performers in the act, the constituent cells, that denotes system behavior. At the end of this course (shortly after birth, when most brain cell divisions have ceased), we are faced with a brain that contains roughly 100,000,000,000 ($=10^{11}$) nerve cells, each of which has an average of 10^4 connections with fellow cells and whose content of about $10^7 - 10^8$ biomolecules (protein) is continuously depleted and regenerated (about 10^4 times in a life span). Thus, if brain action were essentially the outcome of mechanically determined interplay of bouncing and colliding molecules (or cell activities), as presumed by the "atomists" of old, the number of variables (items or "bits" of information) among the elementary component events involved in the construction and operation of the brain would be of the order of $10^{11} \times 10^4 \times 10^4 = 10^{19}$, considering the brain cells as "elements" (or 10^{26} to 10^{27} for biomolecules). The high degree of fortuitous epigenetic variability in the detailed development of each individual brain further compounds this already astronomically high improbability that such an organ could be conceived of as a mechanical machine of stereotyped interlocking gears. And indeed, quite to the contrary, that brain of ours passes through that enormous scramble of microindeterminate performances by its constituent components with the impressive unity and relative stability (or macrodeterminacy) of pattern of which we are personally aware through the sense of persistence of our individual *continuity* and *identity* and the retention of our memory patterns. In other words, far from being a mechanical automaton, that brain is and operates as a veritable natural system with all the features of "order in the gross with freedom in the small" that this term implies.

It will be noted that I am phrasing my statements here in the most general terms, bypassing issues of the mind and consciousness, as well as of conditioning, learning, and other secondary, more machinelike constraints of free systems dynamics, for they would only sidetrack attention from the basic fact that the nervous system, after all, shares the developmental history of the whole embryo and hence must be conceded at least the same degrees of freedom in the sense of microvariance of detail.

It will also be noted that in that cautious version, the assertion of freedom is simply a permissive one, implying no commitment in regard to the mode in which goal-directed choices are made. Let me turn briefly back to the model of the pinball device mentioned before. If set up with the pins in symmetrical distribution, a group of marbles fed in at the top will end up at the bottom in the symmetrical bell shape of a Gaussian distribution, which expresses the fact that every marble, faced at each pin stop with a decision of whether to pass on the right or left, has made its choices at random. Actually, this expression is factually not quite accurate, for mostly a marble will arrive at the next barrier biased by some residual spin from its preceding stop; but for the population as a whole those biases will cancel out. If the pin board is tilted sideways, the symmetry of choices is upset and the population would end up in a skewed distribution, unless the one-way tilt were quickly compensated by a countertilt of equal degree in the opposite direction.

If one is partial to such oversimplified models, one could

visualize any common overall steering of on-the-spot decisions among alternative courses in development as the result of a similar tampering with probabilities within the inner dynamics of the system; indeed, the simile is not without some factual support. However, as soon as one tries to extend it to mental operations, in which the term "bias" has long been quite at home, one hears the question: "Who does the tilting?" My answer, I am afraid, can only be: "The same who has tilted the countless prior processes of development, and all I know about that one is that he works inside the system and not, as I did with the pinboard, from the outside". Never mind that the anthropomorphic question "Who" is wrong phraseology; those versed in systems concepts will understand the answer, while those steeped in microdeterministic folklore will consider it evasive. This is the penalty one faces when one dares venture into the border zone which still separates the content of the life sciences from a full and true understanding of "life".

At any rate, the model has one virtue worth stressing. It circumvents the common aversion of scientists to teleology in the sense of guidance by the future, or "reverse causality"; for the issue fades away if one takes into account that what directs one's choices need not be something in the future, but might well be some physical state in the present; namely, the anticipation and visualization of the goal, represented in the brain as an actual physiological process, or, more correctly, as a specific distortion of the pattern of ongoing processes. Earlier I introduced the mechanical analogy of an elastic net to illustrate developmental system dynamics: any passive deformation of such a net engenders counter forces leading to a new equilibrium configuration. If the visualization of a goal were to produce similarly an actual, rather than purely virtual, deformation of the stress pattern of cerebral processes, this would automatically provide the cues for orienting effector operations toward the attainment of the goal as the unequivocal solution to restoring equilibrium. Imagination would operate by physical imagery.

Some related scheme seems to have been in the minds of the protagonists of classical Gestalt theory. I mention it here

The Order of Life

only to indicate the intimate conceptual correspondence between mental and developmental operations. Thus, what I set forth for the relation of genes to the epigenetic aspect of development, applies *de ipso facto* to brain activity and its expression in human behavior. Here as there, differences between traits can be ascribed to differences between genomes, and lack of a given trait, to gene deficiency; but in a positive sense, no behavioral trait is a pure "gene product", as if exempted from the vagaries, as well as the overall order, of developmental interactions, nor does the genome as such "predesign" the precise pattern of individual behavior. Whether to stress that a given baby is "a striking image of his father" or to be more impressed by the fact that all babies look and act more or less alike, is a matter of personal predilection. In view of the inseparability of genetic and epigenetic aspects of life, it is gratuitous to play up one against the other, particularly in fabricating social, political, legal, or educational philosophies and maxims that highlight either "inheritance" *or* "environment" to the eclipse of the other. Whoever, in full cognizance of the inequalities of genetic endowment, propounds that all individuals are biologically as equal as stamped-out coins, displays just as much partial blindness as he who would view each baby as a uniquely minted coin, having and keeping a full and immutable value designation stamped on it without recognizing that they all belong to a universal currency.

MAN'S MISSION: UNIVERSALITY AMIDST DIVERSITY

Without my having to be more explicit, the reader will understand the allusion to the two divergent antipolar trends in modern society confronting man with precisely those two extremes to choose between for his allegiance and guidance. It is immaterial whether each condescendingly grants to the other "minority status", either stressing an 80% "inheritance" ("nature") and conceding 20% to "environment" ("nurture"), or the other way around; the very failure to recognize their 100% indissoluble, unfractionable, and unending interplay threatens to drive humanity into a state of conflict-breeding social and political schizophrenia. Once again we recognize overstress (now in the psychologic sense of attitudes) as biologically a menace. In short, stress on *individuality*, based on genetic diversity, unless balanced by counterstress on the capacity and responsibility of the individual for developing his individual endowment in tune with the overall, genetically not differentiated, *order of humanity*, is no less lopsided an attitude than is the opposite extreme of stressing the *universality* of mankind, or highmindedly perhaps even of nature as a whole, without placing equal stress on the fact that the only instrumentalities through which human affairs can be executed and brought in consonance with the rule of communal and natural order are the *individuals* in their genetic and racial diversity. The lessons of the scientific study of life—the life sciences—are quite compelling in ruling out as designs for living and survival both aspirations: freedom for irresponsible licentious self-assertion of the individual, on the one hand, and an enforced conformity, on the other, which would deny or abrogate incentive and opportunity for the individual to choose his course for self-improvement by free judicious choices within the range of faculties and limitations set by his native endowment.

Considering that political systems will always tend to veer towards the propagation of one or the other of those nonviable extremes, it becomes a momentous mission for *education* at least to inure the public, present and future, to the uncritical adoption of either, and to do this by soberly expounding the scientific case for approaching controversial issues in the attitude of an impartial jury. The life sciences should accept the mandate to convey to the world, through schools and public media of communication, a balanced view of life as seen from the scientific sector, so as to raise the citizen to a more stable base and vantage point for self-orientation amidst the bewildering bombardment by conflicting slogans to which he is increasingly exposed. In my opinion, this public function of the life sciences is far more fundamental, urgent, and durable than the cornucopia of miscellaneous incoherent scientific news items showered upon a public that wants insight, but gets data—data, moreover, that often enough turn out to be inconsequential and ephemeral, their momentary glitter notwithstanding.

The Order of Life

A modern system of education, likewise, should clear from its offerings in the life sciences, if need be, much of the underbrush of unrelated tidbits of information and stale ready-made conclusions to be memorized; this would make room for the weightier task of introducing the student, i.e., the future citizen, into the *process* of science, the intellectual processing of data into knowledge by distilling, correlating, and evaluating their meanings—meanings not only for science, but for conduct in personal and public life. Just as the developing organism, just as the mature brain, so the educational "system" would thus become true to life in that it would give the student not a stereotyped linear single track to an academic degree ("curriculum"; from the Latin *currere*=running in a track), but an open vista of the objective, which is the conduct of a useful life, of the multitude of ways to get there, of the need to make oriented choices at each branching point with the goal in view, and, most importantly, of the *scientific legitimacy of making choices*.

LIFE SCIENCE'S LESSON: THE PRIMACY OF ORDER

In order to fulfill that mission creditably, the life sciences must themselves practice what they aim to preach. They must not imitate, whether by inadvertence or intent, the political device of seductive slogans which under a pat label say less, but insinuate more, than the plain record of facts and inferences would justify. The fact that general terms, not adequately specified (e.g., "genetic determinism") may arouse connotations far beyond scientific validity, thus leads back to my initial thesis that the impact of the results of scientific research depends in large measure upon the language in which they are articulated.

Obviously, each scientific branch has its own vocabulary and vernacular, not likely to be misunderstood by the "natives". So, why worry? Quite aside from the growing need for interdisciplinary understanding, science has assumed a new obligation vis-à-vis the broad public. Time was when public motivations and acts were guided by beliefs, traditions, superstitions, rumors, and magic. Yet in our day, science is looked to increasingly for guidance, by some merely as a substitute for old black magic; by others, however, as a source of rational counsel. It is on that score that science must rise to new heights of responsibility in making its pronouncements in terms so clear as to render them unfit for abuse as propaganda. The lessons of the discipline of scientific thinking can be one of the prime stabilizers in the unsteady course of modern man. But a misguided scientific following could by the sheer momentum of its numbers lead cultural advance astray.

In the awakening of science to its public function and commitment, the life sciences stand in the forefront. For instance, the problem of expanding populations—chiefly, a biological problem of balance between reproduction and food production—certainly concerns man more directly than does the problem of the expanding universe. But in this new role, the life sciences must conscientiously segregate those lessons which are pertinent to all men as living beings from those which might be used merely to rationalize particular human follies and foibles through superficial similes from nature. Even though no one would, of course, condone uxoricide because, by analogy, some female spiders kill the males after mating, or advocate cannibalism by pointing to the rat mother devouring some of its first litter, the insect superstate of rigid caste organization, in which individuals are frozen once and for all into lifelong immutable stations of forced collaboration, has been invoked as model and justification for certain political systems. Yet, such examples are misappropriated. They go only to illustrate the fate of living species whose phenotypic choices have not had the vastly wider scope given to man in his amplified potential for wise decisions by his experienced and judicious mind. There are more profound lessons to be learned by man from the study of life than to imitate animal behavior.

Yet, where is man to look to learn such lessons? To life science? Is there such a unified preceptor? You note that I mostly referred to life sciences in the plural, which seems to question the existence of something like a unified "life science". Indeed, it does. The disciplines listed as life sciences are still too disparate and often even too parochial to speak with a common voice. Some deal with living forms, others with properties and functions of living beings, still others mainly with techniques of study, recording, and evaluating. Since the sharpness of those distinctions is beginning to break down, one might rightfully ask: "All right, why not try then to extract from them all that essence which all of them must have in common and distill from it a genuine 'science of life', which might even be codified into a sort of counseling manual for man?" In my judgment, this would be neither practicable nor desirable; the open-endedness of present knowledge would defeat the feasibility, and the risk of scholarly science swelling into megalomaniac "scientism" would defeat the purpose. For the time being, we might as well confine ourselves to the more modest task of placing the concentrate of our growing scientific knowledge about life before the people as food for thought—food unadulterated by unscientific admixtures of verbal sweeteners, spices, stimulants, or soporifics—and keep improving the critical ability of people to tell a healthy diet from a polluted one.

Some will choose wisely, others foolishly. This is the margin where the pilotage of the *life sciences* ends and man is thrown entirely upon the steerage of his own *life*. As I said in the beginning, the substance of that ulterior private domain of free self-determination is inaccessible to the objectifying competence of science. What I tried to develop subsequently is evidence that the freedom of choice, which it implies, is fully compatible with scientific experience. The only overall conclusion I dare add, which straddles between scientific evidence and personal belief, is this: *order* is not a secondary result of the ordering of disorder, but a basic principle pervading the universe, although its manifestations are of an unimaginable variety, expressed in phenomena of kaleidoscopic re-ordering and recombining for the emergence of novel forms of higher degrees of order. Just as in architecture the *laws* of structural statics are eternal, but *styles change*, so the basic rules of order are immanent in the universe, and perhaps finite, but they admit of the almost infinite variety of phenomena that we call nature, including man. On this premise, I would feel tempted to extend to nature as a whole the principle of *primacy of order*, which I once enunciated (1940) for the development of organisms: *Omnis organisatio ex organisatione*.

And trespassing where "angels fear to tread", I would even ponder whether "logos" in the beginning of the Gospel according to St. John had not better be interpreted to mean "order" rather than the literal translation to "word", which Goethe's Faust had already tried to transcribe to "sense" and "act"; for the sequence of letters in a word, just as the seriation of nucleotide pairs in the "genetic code", is after all a symbol of order.

EPILOGUE: SO, FINALLY, WHAT IS LIFE?

Like Schrödinger in his essay WHAT IS LIFE? I have thus, at the end, been thrown back on the resources of my own private cognition—wider, but not so universally authenticated and compelling as those of scientific authenticity. Schrödinger asks *"What is this I?"* And he replies: "If you analyze closely, you will, I think, find that it is just a little bit more than a collection of single data (experiences and memories), namely the canvas *upon which* they are collected. And you will, on close introspection, find that, what you really mean by 'I', is that ground-stuff upon which they are collected". This plainly matches the expression I gave in the Introduction to the distinction between private life—let us designate it as L_p—and that part of life, L_s, which can be described and understood in scientific terms, culminating in the inequality: $L_p > L_s$.

However, I have let the scientific propositions about principles of living systems lap over farther into the private sector than has Schrödinger. I did so in promulgating the thesis—scientifically supportable—that the personal experience of "freedom of will" has a legitimate scientific counterpart in the decision processes of "adequate choice" inherent in the "microindeterminacy" of natural systems. Therefore, when Schrödinger (p. 88) says: "I think that I—I in the widest meaning of the word, that is to say, every conscious mind

The Order of Life

that has ever said or felt 'I'—am the person, if any, who controls the 'motion of the atoms' according to the laws of Nature", the scientific 'I' of this writer feels justified in amending the statement as follows:

The "I" does not control the "motions of the atoms" and molecules *directly*. It can affect them only through *the agency of intermediate subordinate systems, ranged in a hierarchical scale of orders of magnitude*; the pattern on each level being macrodeterminate (with "order in the gross"), the lack of stereotyped microdeterminacy in its effectuation ("freedom in the small") notwithstanding. This formulation, based on the broader scientific study of the mode of operation of living systems in general, subsumes, as one can readily see, the feature of "goal-directedness" of the decision processes of our private experience.

It seems to me that this extension of the posture of Schrödinger marks real progress during the last quarter century in the rapprochement between the subjective "inner" and objective "outer" aspects of our lives. Rather than trying to squeeze our self-knowledge into the all-too-narrow frame of contemporary scientific knowledge, true science, fortified by logic, tends to expand its own scope so as to enlarge the area of mutual correspondence. In this sense, *science has not only left inviolate the primacy of our own cognizance of "freedom of will and choice" but it has given it a scientific underpinning.*

POSTSCRIPT: FROM LIFE SCIENCE TO EDUCATION FOR LIVING

EDUCATION: GUIDE TO HARMONY

As a composer would preview the main themes of his opera in the overture, so many scientific journals have adopted the style of having the Summary of an article precede the text. A Foreword usually has the same anticipatory function; which does not mean that the prognostication will be fulfilled. This postscript is to supplement the unfulfilled function of the preceding text.

The key theme of my Foreword has been a call on science to help man break out from his self-spun cocoon of parochial self-confinement and rise to heights from which he can view and appreciate the unity amidst diversity of the world in broad perspective. Yet looking back at my treatment of the topic of this book, I recognize that, however broad it be in scope, it cannot, taken by itself, be more than a pointer—a roadsign for mental reorientation away from ruts of traditional preconceptions toward a fresh and critical look at nature and the sources of our knowledge about nature and ourselves. It can be no more than, at best, part of a blueprint for a building program, without the building process having hardly started. And even as a blueprint for action, it is incomplete, for it is more explicit about how to dispel misorientation than it is about positive guidance. Thus, to correct this imbalance, I have felt compelled to add the missing coda in this postscript.

The motto given in the Foreword for the book was "to substantiate the scientific rationale and validation of civilized man's obligation to strive for a realistic and balanced perspective, in which he recognizes ideological extremes for what they are: artificially disconnected opposite ends of continuous scales of intergrading values, just fortified in their antagonistic isolation by verbal symbolism and the instinctual vestiges in man of his biological past. The example of the reconciliation of the rule of order in nature with the legitimacy of freedom and diversity serves merely as a model in point. It, so to speak, sets the tune".

The Order of Life

My effort to reconcile, or rather merge, putatively incompatible alternatives—holism and reductionism; determinism and indeterminacy; field dynamics and particulate autonomy; field—gene "solipsisms";—were meant to serve as illustrations of the larger task ahead. But as for the implementation of that task on a universal scale, I abdicated to wishful thinking in these words: "If *education* would pick up the tune and amplify it, and if it were to find the proper resonance in men's minds, it would transmit to man a cultural gift from science." In short, unless followed by action, mere counsel for action is a futile undertaking. Thus, if I called on science to provide wise counsel, I placed the call for effective action right at the doorstep of *Education*.

This is no longer an esoteric academic matter, but one of grave concern and urgency in the contemporary scene of bewilderment, disorientation, and erratic fumbling of a supposedly educated and civilized citizenry. Has education grown too diffuse or has it left no durable residues on the mind? The vacuum it leaves behind is fast being occupied by *indoctrination*. A population, craving increasingly for absolute security of body and mind—the lethal soporific administered by soothsayers unmindful that moderate stress is a biological prerequisite for health and viability—seeks intellectual shelter behind "iron-clad" pat answers to questions that otherwise they might have to decide for themselves under the strain of mental effort involved in critical decisions. Why else would there have been such signs of relief when the sheer faith in fatalism in the public mind was given what seemed to be scientific approbation by the misconstrued term of "genetic determinism", wishfully translated to: "If every act of mine is preordained by my genetic blueprint—and note, not just as the frame within which I am free to chose, but as the positive manipulator of my actions—then why should I submit myself to the stressful efforts of conscientious judging and striving?"

Of course, I am not accusing science of such inane perversions of its dictums. Yet, it seems timely to put scientists on guard against the potential abuse of the content of their Pandora's Box of statements apt to become innocently converted into pernicious, though palatable, fare. Never before in history has human reason had to contend with such powerful devices for the amplification and potentiation of untruth, half-truths, and confusion, as the media of public communication are capable of showering upon an unprepared, uncritical, half-innocent, half mischievous, part-literate, part-barbarian mankind that has not yet been immunized against the intellectual pathogenic microbes, unwittingly or insidiously let loose by that barrage of catch phrases. In searching for an antidote to that endemic, an effort to "detoxify" misinformation and misinterpretation would hardly be a promising approach. *Immunization* at an early impressionable age would. And this is where a Liberal Education, unwrapped from the mothballs of the past and re-invigorated by the mandate for a new vision of the future of man—the literal meaning of "humanism"—must come into the act. It must arouse man from his monomaniac concentration on momentary or endemic "predicaments" to an attitude of cultivating his rational endowment for moving on, eyes to the future, from his illusory notion of a world in status quo to a truly humane status, soberly reconciling his creative urges with the limitations set by nature to feasibility.

To live up to this ideal, education will have to put on some sort of bifocal lenses, as it were—short-range for dealing with a contemporary world in distress, and long-range for the charting of prospects for a less unbalanced future. This dual purpose defies any attempt to set the goal of education up as an either-or proposition favoring one alternative over the other—for instance, either stuffing the student full of data of information and practical tricks of the trade through which to "make a living", or rather widening his perspective and understanding of the world so as to make the living that he will make, really worth living in terms of "enhancement of the quality of life", as the current slogan goes.

Unfortunately, many recent well-intentioned, but one-sided, plans for educational reform are doomed to failure by the schizophrenic compulsion that makes them formulate the problem in such a fallacious "either-or" dichotomy. The issue of education thus joins neatly the family of sham an-

The Order of Life

tagonisms which I tried to unmask as such in this book—holism vs. reductionism; fields vs. particulates; etc. Similarly, education argues about: vocational efficiency vs. erudition and judiciousness; memorized robotry vs. intelligent self-direction; short-range cures for mankind's current ills vs. long-range planning for the prevention of future recurrences; etc. This dogmatism can serve as an introductory illustration of the kind of roadblocks education will have to raze in order to clear the road to a saner future for man. Let me start from the short-range aspect.

"LINEAR CAUSALITY": PREDICAMENT OF MAN?

Bemoaning what is called the "predicament of man" has become a popular pastime. It comes in a scale of versions from the deeply serious concerns of the global governmental and private agencies, who have enough crucial information on hand to know the background of the symptoms of the predicament, through the excited and well-motivated, but not always well-informed, amateurs, down to the devotees of parlor games with stakes no higher than craving for attention. Attitudes vary just as widely. At one end are the optimistic boosters, who forthrightly deny that the world is creaking in its joints from growing pains, and at the other end, the congenital pessimists, prophets of doom, proclaiming that the world has already become disjointed past the point of possible return to health and sanity. The retort to both extremes is that the world is, after all, like all organic nature, in the process of evolution, but that its course and fate, while unpredictable, are left, for better or for worse, to the wisdom or whims of *man*. Man can and must act. The do-nothing attitude both of the optimist, stemming from his faith in preordained "progress", and of the pessimist, from his obsession with futility, is the only attitude with a safely predictable evolutionary outcome: dissolution, decay, extinction.

Thus, steering clear of both extremes, most people would chime in with the call for action. But what action? If one looks at the present scenery without prejudice, one notes that most pleas are for counteraction; yet, counteracting what? Evidently, in every particular instance that which the particular group has picked out from its limited sectorial angle of parochial vision as the presumed deleterious "cause". And if one tries to add up the various antidotal nostrums proffered, one finds that they have been so diverse and contradictory that their net sum lies in dead center—consummate ineffectiveness. Collectively, they are as ineffectual as the "do-nothing" scheme.

In this state of impasse, one would first explore what solid substantive content one could distill from the somewhat nebulous generality "human predicament". This exploration has led me to the conviction (1) that it is useless to look for whatever lies behind that so-called predicament of man until the phenomenon itself has been more clearly itemized and specified; (2) that having done so, we shall find the phenomenon to be not at all so inexorable as the term insinuates; (3) that to make this hopeful prognosis come true will require dedicated effort and a spirit of "realistic idealism" (see later); and above all, (4) that the medication, or rather the prescription, for a healthier and saner world lies not in contriving fanciful new ideas on how the world should be managed, but rather in showing to man convincingly how he has mismanaged his affairs, in order that he may, of his own account and from his own intelligent reasoning, stop that malpractice. In short, in contending with his predicament, man's mind should be turned less on what to do than on what *not* to do.

As I propose to elaborate, this object is far from a "do-nothing" precept. It is something that requires massive action—"massive" by momentum, not by multitude. Whatever number of incoherent symptomatic remedies to this or that headache of mankind one might advocate, they could, at best, palliate malaise but hardly add up to an effective preventive of future disease. However grave man's concern about a given crisis be, and however noble his preoccupation with efforts to stem it, a far weightier task ahead is to discover how to prevent his past practice of crisis-hopping from continuing unabated into the future. And this objective, I maintain, cannot be served short of a profound sanitation of man's traditional *frame of mind*—a cleansing operation that clears out accumulated misconceptions.

The Order of Life

It is remarkable how much confusion, misorientation, and outright perversion of man's thinking and actions, leading to human strife and misery, can be traced to the misunderstanding, misinterpretation, and eventually, misapplication of valid knowledge. *Mislabelling* lies at the root of it all; for most of the terms in which we deal with knowledge are shorthand labels for things, processes, relations, concepts, and so on. By force of habit, man is prone to adopt heedlessly a seductive brand-name on the label of a bottle as substitute for a description of the content. Our terms of language, from everyday vernacular to scientific idiom, can be just as insidious; I have referred to this trend by illustrative samples in the main part of the book.

Among the commonest of such labels is the term *"cause"*, if it is left unqualified. I shall not dwell on its historical roots; they are, I would submit, deeply embedded in man's extrapolation to nature of his own spontaneity in willing an act. Presumably, primitive man then went on to populate the universe in his imagination with "actors" after his own image. Sophisticated man simply reversed the process by invoking "primary causes", which he then let be followed in linear ascending series, domino-fashion, by secondary, tertiary, etc., causes, confident or persuaded that ultimately the causal chain will "explain" man's own spontaneity, which had served him as the model for the whole argument in the first place.

Being inexpert in the history of ideas, I cannot vouch for the correctness of this origin of the concept of "linear causality". Yet, the circularity of logic in its conception seems to me obvious. One need only compare the term "cause", as it occurs in the naive, unqualified usage of everyday language, with its true content of meaning, as it presents itself to sober critical examination. The Oxford Universal Dictionary defines "cause" as "That which produces an effect", and "effect" as "something caused or produced". Evidently, quite aside from the fact that this circular language leaves both terms unexplained, as neither is reducible to the other, the anthropomorphic reference to "production" is a tell-tale recourse to animism and spontaneity. To indicate the range to which this "anthropomorphization" of nature has expanded, let me just point to the extension of the legal concept of "responsibility" to the interaction of phenomena in the inanimate world, such as: "A bolt of lightning was 'responsible' for the fire"; or, "a tidal wave was 'responsible' for the destruction ashore"; and so on; "responsibility for. . . ." being used as synonym for "cause of. . . .".

Now, just stop to ponder for a moment what we really try to express by those phrases. Let me be quite elementary and consider a simple "causal chain": the turning on of an electric light by throwing a wall switch. My muscular energy (a) moves a piece of metal into the gap of an interrupted electric circuit, thus "causing" (b) an arrested current to start flowing to the bulb, and there the energy of the passing current "causes" (c) the cold filament to heat up to incandescence, that is, to a degree of energy emission which when it strikes our retina, (d) "causes" the optic nerve to send impulses to the brain which (e) "cause" the sensation of light. This would be the ordinary shorthand wording to which the enlightenment of our civilization has gradually converted the much simpler explanation by mystery and magic with which a primitive savage, watching the performance for the first time, would have been satisfied. In both cases, the urge to simplified expression testifies to the parsimony of the human mind. But has not, in our own case, sophisticated "analytical" oversimplification perhaps led us astray by imputing similar simplicity to the natural phenomena which we aim to relate? Is our description of the passage from turning on the light to seeing it in terms of a linear sequence of isolated steps really representative of the way nature operates?

The truth is that to view and treat such a "causal chain" as if it were in fact reeling off in isolation and sovereign independence of the web of the dynamics of the rest of nature, in which it is enmeshed, is a result of habitual imagery. It rests on an abstraction that has been eminently useful in giving us a coarse-grain effigy of nature for our practical guidance, but fails us when we are after deeper and comprehensive insight. I would compare its utility to that of the Mendeleev table of elements, in which whole numbers were assigned to atoms, until the critical departures from that simple scheme called attention to the reality of isotopes.

In a similar sense, the "causal chain" concept is applicable to given circumscribed tasks (or problems) for the attainment (or solution) of which it is quite irrelevant that the concept has only approximate, but not absolute, validity; that is, to situations in which consideration of the total context in which the chain is thought to operate proves to be empirically dispensable to all intents and purposes. To recognize this limitation, just note how our sample sequence of switching on a lamp, $a \to b \to c \to d \to e \to \ldots$, would have lost its linear simplicity if we had given thought to the deliberately ignored and bypassed contingencies of a broken electric switch, a power failure, a burned-out light filament, or blindness of the reporter. Of course, for each one of those contingencies, we would have construed its own causal underpinning. But tracing matters back this way evermore backwards, we find ourselves either receding ad infinitum, or arriving at an ultimate stop sign—some verbal term coined by man as invocation of a symbolic "ultimate cause", not unlike that which primitive man had reached by a mental shortcut. In both the simpleminded interpretation of the latter and the more sophisticated version of our own age, the principle of linear causality lacks comprehensiveness and pertinence. As I have stressed throughout this book, we should no longer conceive nature as a bundle of causal chains but as a hierarchy of systems operating according to the rules of multivariant network dynamics, in which certainty has had to give way, in some measure, to probability.

This development in no way implies the renunciation of causality as such, but it disavows the dogma of causality being executed by linear chains of cause-effect sequences. In fact, a critical examination of what the term "cause" truly connotes, or rather what it does not imply, will rescue the general thesis of "causality". from being thrown into discard along with its chief exponent, "cause". As far as I can discern, "cause-effect" relations have purely *negative* connotations, as follows. We focus our attention as a matter of deliberate abstraction on a given fraction or "Subsystem" (s) of Nature (N); we thus oppose s to ($N - s$), which we shall call E; we then register changes of s between times t_1 and t_2 from state s_1 to state s_2 ($\triangle s$); concomitantly, we record correlated changes in E, ($\triangle E$), which we may have experimentally produced or statistically established. We further note that $\triangle E$ (and hence, E) contains two compartments, of which only one, let us call it E_V, participates in the change $\triangle s$, while the other $(E - E_V)$, which we shall call E_C, appears to remain uninvolved or constant ("silent") in that particular change of s, on which we had focussed. Conventionally, we then go on to designate $\triangle E_V$ as the "cause" of $\triangle s$, unmindful of the fact that we have arbitrarily disregarded the integral connectedness of E_V with E_C, i.e., the essential dynamic continuity of nature. Objectively, all we are entitled to conclude is that a given *change* in the state of s (the "effect" $\triangle s$) would *not* have occurred in the *absence* of the *change* in the state of $N - s$ (the "cause" $\triangle E_V$).

Following this argument, a "cause" reveals itself as being simply a definable *change* of something in nature (of a state, process, configuration, etc.) *without* which an unequivocally correlated *change* in nature, the "effect", would *not* have occurred. This formulation cannot be turned around into its positive inversion, as is done by submitting that a "cause" positively specifies a given set of "effects" without taking fully into account the whole setting. In short, our mind dissects, as it were, those features that have changed from those others of the total cohesive context that have remained stationary, and it ignores or forgets that in reality changes of a thing cannot be separated from the thing itself. Therefore, the native mind, which has come to regard "causes" as autocratic actors with generative or creative powers, comes to realize by degrees that the word "cause" is in essence only a personification of *differentials*—of the differences, variations, and modifications which a given *existing* state of nature undergoes. The personification lies in vesting attributes with the substantive status of subjects, for we present and treat differentials as if they were independent of the entities whose variations they signal (see the discussion in the main part of the book regarding gene-controlled "characters").

To sum up our conclusion in terms of the symbols used above, it is crucial to remember that in establishing a causal nexus between changes in a system s and changes in $N - s$,

The Order of Life

the presence of the "silent" component E_c of the latter may be taken for granted, but must never be left out of consideration. It is true that in practice we confine the scope and time span of our attention to such narrowly circumscribed samples of the dynamics of the world that within those confines any variance of E_c appears as practically negligible compared to the appreciable changes of the component E_v which we have singled out precisely for that reason, guided by empirical prescience. But let us always bear in mind that the validity of this artificial separation of E_v from E_c is limited to the range within which we have found it valid by experience and empirical tests. We must not let this expedient obscure the basic fact that, in principle, nature (N), and hence, $N - s$ (=$E_c + E_v$), remains an unfragmented, indivisible, dynamically continuous system of integral unity whose relevance to every phenomenon in the world must never be discounted, even when it may be disregarded in specific simple situations.

In the light of the preceding discourse on causality, it will have become self-evident that decisions reached in the artificial self-confinement to linear cause-effect considerations, with the attending neglect of total context, background pattern, circumstances, or under whatever name we wish to allude to the intrinsically systemic character of nature, are *scientifically* of questionable authority. (I refrain from considering the legal counterpart of the argument.)

To realize this fact is of prime importance if human thinking, which underlies the attitudes and actions that contribute to the "human predicament", is to be set straight. In our world, in which scientific verdicts or opinions have become noticeably influential in the moulding of public attitudes, the scientist must exercise increasing care lest his dictums contain ambiguities or exaggerations that might lend themselves to exploitation and abuse in the interest of political partisanship. The simplicity and convenience of uncritical submission to the autocracy of the principle of *linear* causality has conditioned public thinking to such political seduction. And since science has had a major share in the past certification of that principle, the scientist, having now recognized its limitations, has a social responsibility to immunize the public against the hazards of its continued misinterpretation in the future. This sentiment has motivated me to supplement my earlier disclaimers of microcausality by the following more explicit demonstration of social risks inherent in the thoughtless extension of classical causal thinking beyond its legitimate borders.

CATEGORY VERSUS CONTINUUM: CONCEPTUAL CONFRONTATION?

People brought up on strictly linear causality view the world not as a *continuum*, but as a *mosaic* of discrete, neatly delineated entities, linked in the functional asymmetry and polarity of one entity (functioning as the "cause") being the active member in a process, whereas the other entity (the "effect") follows passively. This fragmented picture of the world is reflected in the human habit of pigeonholing mentally the pieces and aspects of that world mosaic in sharply bounded categories, such as molecules, sounds, drugs, races, peace, freedom, and so forth.

Now, there is nothing wrong with classifying natural phenomena for convenience by placing them in groups according to major sets of attributes or criteria shared by the collective regardless of the minor differences that distinguish the members of the class from one another. Our ability to identify such categories in nature is the mental corollary of the hierarchic systems structure of the universe, a category having the same features as has the natural system that it categorizes, namely, stability of overall pattern combined with variety of detail. It reflects what I have called the rule of systemic order in the gross, coupled with individual freedom in the small.

Yet, serious misconceptions arise if one forgets that categories have only statistical validity, that they denote only the *average* aspects of a given group of phenomena, but are wholly devoid of any *determinative* meaning regarding the particular properties and states of the *individual* constituents of the collective. They specify only the range within which the latter lie and behave, but are strictly non-committal about

The Order of Life

the specific places and performances of the constituent units of the system. Moreover, like all systems, natural categories are, in defiance of the sharpness of their verbal definitions, always fuzzy at the edges. Thus categorization has both its merits and its limitations. And to transgress those limits, whether in mind or speech, spells trouble.

Disregard of this warning keeps swelling the pool of emotional and political pollutants that pervert potential human enlightenment into manifest human predicament. One need only recall the nonsensical and pernicious habit of many people to prorate a collective appraisal, which is perfectly valid for an assayed group as a whole, to the individual members of the group, as if they were stereotyped, all alike, each one a fair and representative sample of the group. This logical atrocity, while grist for the mill of unscrupulous politicians, is of course a horror to men steeped in critical thinking.

But do we do all we could to help critical thinking assert itself in the public mind over the blind uncritical acceptance—and subsequent amplification—of other peoples thoughts? This is where education would have to step in. All it would take is the repetitive exposure to a good selection of pregnant examples that illustrate the absurdity of viewing, and hence, treating, the mass of members of a collective (or constituents of a system) as a homogenate. No child would miss the point if pointed to the fact that if mortality from a given disease is, let us say, 10%, this still does not mean that any given person contracting the disease will die 10%, for naturally that person will end up either 100% dead or 100% alive.

To neglect the intrinsic inhomogeneity of the populations of items that are subsumed under a single categorical designation is only one of the aberrations of logic which a shift from the tidiness of microcausality to the wider tolerances of system thinking could rectify. A second one, equally devious, is the deterministic notion of the fixity of content and contour of categories, to which champions of microcausality must resort. Their scheme demands that categories be rigidly delineated against one another by absolutely sharp boundaries, admitting no latitude for graded transitions, like the political artifacts of national frontiers and of party lines in elections. For this frame of mind it becomes axiomatic and a matter of course to reject the lesson of experience that in nature the dynamic interactions among systems always blur the imaginary hairlines of categorical distinctions. And even where graded transitions are plainly evident, they are ignored or presented in the light of barriers rather than of links, all for the sake of attaining maximum dialectic contrast. It is evidently up to education to show how this accent on *confrontation*, on the insistence of formulating complex problems in the simplified dichotomy of black-versus-white issues by the deletion of the connecting gray shades, has yielded many of the familiar antithetical postures that have barred reasoned conciliation of age-old conflicts.

Perhaps it is in this place that the scientist—not the "nose-to-the-grindstone" specialist, but he who has not yet lost universalist communion with nature—has a major commitment to fulfill to education. My reference here to the scientist as a *human individual* knowledgeable about nature, rather than as high priest in the worship of an allegorical abstraction, called "science", is deliberate. It tries, in the spirit of my general call for blurring the demarcation lines of such categorical distinctions as that between "science" and "non-science", to dash the illusion that science holds the master formula for the acquisition of balanced attitudes in the judging of issues. "Science" can do no more in this role than help man straightening out his somewhat warped picture of the world. It can rid him of some of his insidiously ingrown fallacies by showing him the beautifully harmonious continuity of nature to replace in his thinking the simplistic, coarsely granulated and distorted image of nature, full of cracks and contrasts, with which his animalistic past has encumbered his mind.

Man's proclivity for thinking in terms of polar extremes is a residue of that past, and he likes it. Reducing problems to the simple absolutes of the categorical antitheses of the "black-white", "good-bad", "right-wrong" variety lulls him into the comfort of a fairyland of certainty and intellectual security, which he craves. By disillusioning him in these primitive expectations, science can point him from his past to his future destiny, from his evolutionary past of fumbling progress by trial-and-error, to his destiny of a future of rationally willed goals, based on insight and foresight. In calling on him to use his rational faculties which would enable him to defuse the predicament-breeding charge of divisive *polarization* by deliberate *centripetal* and convergent orientation to what humanity has *in common*, science is only joining, on rational grounds, the voices raised by others on ethical and moral grounds in behalf of the same objective.

TOWARDS DEPOLARIZATION: TASK FOR EDUCATION

Yet, without removing the handicap of the epistemologic adherence of public thinking to the outdated concept of nature as a precision machine, converts to the adoption of the spirit and attitude of *"depolarization"* would come too haltingly to form a crucial antidote to the spreading malaise and overt ailments lumped under the term of "predicament of man". The required abandonment of the smug sense of security, which faith in causal microdeterminacy instills, in favor of a certain measure of uncertainty inherent in systems thinking; the acceptance of higher responsibility for self-determination and self-direction which this implies; the sense of giving up categorical guide posts, supposedly well defined and delineated, for an uncertain promise of a "better world" of communion; the mollification of human cussedness, deprived of the emotional satisfactions attending strife; all these are sacrifices that will not be readily assumed, and least of all by a population brought up to believe in a preordained fixity of the world, with change being just an inescapable aberration.

But what about future generations not raised on those static fallacies, but inculcated from childhood on with the realization that in an evolving universe, from the celestial bodies down to organisms, man included, it is not statics, but *dynamics*—change—that is primary and inexorable; that man has the faculty, and indeed the moral obligation, to influence that change for the better within his own orbit; that the exertion of such an influence is a matter of give-and-take, for in a thermodynamic economic world every move exacts a price; that reason dictates to minimize the price; and that acting on conventionally polarized rationales is far more costly, to the point of bankruptcy, than reasonable depolarization would be? Would such a generation not have less hesitation about choosing salvation in preference to perdition, even at a price?

Since I believe that it is high time for us to start preparing the ground for such an unspoiled generation to thrive in, I would propose to entrust the implementation of the task to all those who consider "education" still as the *raising of a civilized human being*, and not just as the imparting of information and training for a job. In that sense, education starts at home, and since the public media pervade home life with ever greater volume and impact, they could perhaps take part in this change of course away from abiding by the animal trend toward "polarization" to advocating *the human need and potential for "depolarization"*.

These ideas are not so unrealistic as they may sound. They do not submit, of course, that a simple about-face in educational outlook and policy would succeed in switching human preference from habitual antithetical polarization of issues to conciliatory resolution. But what education could achieve is sensitivity for recognizing and rejecting unreasonable artificial accentuation, and hence, aggravation, of the polarization of issues, which in broader perspective would reveal no aspect of irreconcilable antagonism to an unspoiled critical mind. At any rate, I labored the either-or attitude merely as one of the many illustrations of traditional attitudes that ostensibly contribute to the bemoaned "predicament of man"—not to mention the fact that the "predicament" has not arisen in its full magnitude all of a sudden, but only has surfaced or been brought to the surface of perceptibility by byproducts of our civilization, which tries to grow faster than our civility.

After this floating sample exercise, I want to move on now to the firmer ground of the practical counterefforts education might undertake in extending concern for public health to the area of human *thinking*, the *language* which expresses thought, and the *behavior* that translates it into action. Yet, even this extension will require a thorough rectification of the theoretical and tactical flaws which I detect in the thought structure on which our conventional education rests.

EXAMINING THE EDUCATIONAL PROCESS

I have never been able to view the educational process otherwise than as an indivisible continuum, rather than as the succession of discontinuous compartmentalized events into which practical necessities of "school systems" have chopped it up. To realize the degree to which this conceptual fractionation of the educational continuum has proceeded, one need only consider the current wave of efforts to cope with the very real inadequacies of "University Education" by puny tactical administrative nostrums, instead of striking at the heart of the matter, which is the lack of basic clarification and formulation of what the *purpose and aims of all education* in the modern world are—or rather, ought to be. When students of development, whether embryonic, mental or cultural, or of history, who are fully cognizant of the uninterrupted progress along the dimension of time of the processes they study, nevertheless indulge in breaking the continuous course up into stages, phases, epochs, or what not, they do so for laying stress on certain prominent focal features characteristic of those periods, without insinuating that there are sharp demarcation lines between them. Yet, in administrative procedure, the areas of transition between such domains of emphasis have a way of freezing into artificial rigid boundaries, not unlike national frontiers.

In order to counteract such adverse trend toward disintegrating the conceptual continuum of education, albeit in full appreciation of the practical necessity of packaging it for convenience into artificial fractional boxes, one ought to develop a set of core principles, and perhaps codes, that apply to the process of human education *in its totality*, "from the cradle to the grave".

(1) WHAT IS THE PURPOSE OF EDUCATION?

In this simplistic form, the question is too indefinite to lend itself to any but the tritest answers, such as, for instance, "to impart knowledge". The question might gain in precision if we were to identify education with the formal transfer of information. But in this restrictive sense the question would be grossly misleading; for it would present the human mind as essentially an empty receptacle to be filled with extraneous "input", ignoring the continuous intrinsic processes of growth and maturation of the mind. To those who consider mind as a matter of strictly personal *experience*, the question of whether or not it has a prenatal history might seem to lie way beyond the testing range of objective science. Yet, the lessons of modern life science have incontrovertibly demonstrated that just as the indispensable instrument of the mind—the brain—has, in common with all organs of the body, its unique individual ontogenetic course of morphological and functional differentiation, so do all the overt manifestations of its activity, which we call "behavior" and from which we infer the mental activity of other persons. Accordingly, when an individual is born, it has already gone through a long prenatal phase in the formation of its characteristic and individually unique think-structure. It is important, therefore, to recognize that the shape of this native endowment circumscribes the scope within which the conscious mind can operate.

Within those boundaries, the space left for the finer articulation, perfection, and maturation of the mind is immense. However, the specific manner in which this content of the mind is progressively enlarged, refined, and further elaborated into thought patterns, must be viewed as essentially an adaptive *modification* of the pre-existing prenatal pattern, rather than as a totally new *addition.* In this sense, the doctrine that "all men are created equal" is, of course, scientifically untenable regardless of whether we compare them at the moment of conception or the moment of birth. Each individual's native endowment is his or her inalienable private distinction. Although the range of opportunities that it leaves open for self-development is considerable, it is by no means infinite. It is definitely limited by the repertory of capacities potentially made possible within the native endowment. Consequently, any educational system that would fail to take into account this inequality of potentials among persons would be building on quicksand. Incidentally, it will have been noted that I have not described the "native endowment" as "genetically programmed" in the conventional

The Order of Life

sense of "gene-determined"; for the unpredictable variety of "epigenetic" circumstances, to which the developmental process is exposed, rules out any such notion of absolutely preordained fixity (see the discussion of evidence in the earlier sections of the book).

In conclusion, the only realistic answer to our initial question of "What is the object of education?" would have to be "to help the process of *self-development* of each individual reach maximum proficiency attainable within the potential of his or her capacities and the inescapable limits of personal faculties". This, incidentally, is precisely the original meaning of the term "education", which connotes "to bring out", and not "to stuff in".

(2) SELF—DEVELOPMENT AND UNIVERSAL EDUCATION: ANOTHER CONFRONTATION?

Let me at once call attention again to the misleading form in which this question is often posed. It uses the age-old political trick of polarizing issues into alternatives of choice between two purportedly mutually exclusive opposites instead of striving for optimum mediation along the graded scale of values that connects those extremes. If humanity is to mature to the sanity of moderation, how could it get there as long as the educational process, supposed to prepare man for the future and thereby to prepare the future itself, indulges in the same vicious habit of putting into categorical opposition the extreme ends of continuous scales of values?

I am using the phrasing of my second question merely as an example of the kind of phraseology which, if one comes to think about it, is educationally pernicious. Let us just briefly consider why. My postulation of *individual self-development*, according to point (1) could in the combat between feuding educational philosophies easily become exaggerated into a claim for absolute self-expression with addiction to all kinds of personal whims and monomaniac withdrawal from all restrains of self-discipline. Contrariwise, the mere mention of *optimum education for everybody* could readily be played up by the opponents of egalitarianism as a clarion call for a procrustean ideal of overall mediocrity—"procrustean" in the sense that the over-endowed would be forcibly stunted, while the under-endowed would be artificially stretched, all to a common level of the average, unmindful of Aesop's fable of the goose which, however much it might stretch its neck, never succeeds in turning into a swan. And so, before we know it, we shall have the partisans of the two extremes confront each other in battle lines, their respective battle cries, of course, being "absolute freedom, indeed unbridled license, for the individual" and "absolute supremacy of an egalitarian social conformism".

In contrast to the futility of interminable contest between such emotionally satisfying, but intellectually absurd, antithetical propositions, man must learn to substitute for the simplistic *either-or* form of alternatives an exploration of what *combination* of both would in proper proportion yield the optimum solution to a problem. Although that proportion would vary considerably with cultural conditions, societal mores, political philosophies, economic circumstances, etc., and at any rate, will fluctuate in time, it must be anxiously guarded against ever shifting too far toward one or the other extreme.

To some extent, the safety margin for proper balance is dictated by some basic verities. In the first place, the *common stock* (C) of properties, features, faculties, and limitations that are shared by all human beings, is far greater than the *personal distinctions* (P) among individuals ($C \gg P$). This natural disparity already gives us a key for the relative apportionment for educational attention and emphasis. In addition, however, it implies a practical obligation which, to my knowledge, has never been made sufficiently explicit in any educational system, namely, the obligation to recognize the legitimacy of the principle that decisions such as we are discussing here can never be made in absolute and apodictic terms, but must vary substantially depending on the context in which they are made. In other words, with due regard for the need of cultivating both the right of freedom of the individual for self-developement, on the one hand, and the restraints which his belonging to the higher collective of humanity and to its societal subvariants impose upon him, on

The Order of Life

the other, we must recognize that in real life each individual is bound to be confronted many times with the conflicting counterpressures arising from the two kinds of discordant propensities. I believe that it is up to the educational process to expose everyone early in life to the unpalatable fact that whenever he or she may be called upon to act, they must let their decisions be guided not by a ready formula in favor of absolute egotism *or* absolute humanitarianism, but rather by judging on their own responsibility in which direction to weight the decision in consideration of the *context* on which the action to be taken has a bearing. Identical sets of situations acquire totally different *meanings* in different contexts and accordingly must receive different ratings in the decision-making process. I shall return to this point further below. For the moment, let us just think of the wisdom in the popular phrase that says: "It all depends".

So, in conclusion, not only can there be no global formula of where to draw the line between promotion of self-development and awakening to the need for self-restraint in deference to collective order and loyalty, but education must condition the individual to accept this degree of indeterminacy, and hence, of self-determination.

(3) FURTHERING PERSONAL SELF-DEVELOPMENT

In medicine, therapy must rest on proper diagnosis. But the latter must also include an estimate of what share in the recovery process the powers of resistance and healing capacity the organism itself is likely to be able to muster. The educational counterpart of this is, of course, the identification of personal characteristics, comprising aptitudes as well as deficiencies; this is the sort of thing vaguely referred to as "personality structure". One might cite under this heading such componental features as manual dexterity, intellectual acumen, motivational strength, perseverance, imagination, breadth of interest, critical judgment, abstracting power, and so forth. Let us not overlook, however, that all such listings represent artificial fractionations of the mind and that the various components here singled out are not only all owned by every individual, albeit in widely varying proportions, but are intimately inter-related among one another, in some cases in mutual re-inforcement, in other combinations in mutual competition. This being the case, the educational practitioner, comparable to a one-sided clinical specialist, might want to ask himself which ones of those personality traits to favor especially in their development at the expense of neglecting some of the others. Should he concentrate his and the individual's attention on those with promise of becoming truly outstanding or should he, on the contrary, direct his primary efforts at enhancing the stunted growth of the constitutionally under-endowed traits?

You note that the dilemma is not unlike that met in political decisions about how to distribute most propitiously limited funds and energies in aiding underdeveloped countries or under-privileged groups within mixed populations. On the basis of my biological experience, I must again insist that posing such questions in "either-or" terms can only lead to strife, instead of rational solutions. There is no property of any living organism, structural or functional, the contribution of which to the efficient performance and survival of the whole of the body could be evaluated without consideration of the rest of the whole system. Consequently in ignorance of the intricate web of dynamic interactions among the many partial processes and functions of a growing system, it would be impossible to predict whether single-tracked preference for special attention given to any one set of features would bring *summarily* harm or benefit to the completed system. In its application to personal traits, the uncertainty in this "cost-benefit ratio" mounts even more if one takes into account the duplicity of benefit judgments depending on whether they refer to the individual or to society.

So, here again, our answer is a call for mediation between two extremes, aiming at giving full scope to the self-development of special talents, but pari passu strengthening, rather than allowing to atrophy by disuse, the weaker potentialities of the personality, the optimal harmony of which makes human character.

I realize that I am touching here a very "touchy", and certainly politically sensitive, point in implying that basic education should be of the "intransitive" kind, helping the individual to develop primarily into a well-rounded being, fit to lead as useful and harmonious a life as his native endowment would permit, rather than of the "transitive" type envisaged as education "for something special"; this latter being hardly education at all, but training. In making this distinction, I have in mind, of course, the potential conflict between the capacity and urge of an individual for effective and satisfying *living*, being given strong vocal expression by modern youth, and the necessity of the individual to *earn* a living by engaging in tasks and services—"jobs"—increasingly demanded by a rapidly differentiating society and allegedly rewarded by it in some proportion to the merits of performance. This is where vocational and professional *trainings* for special competences far beyond a general background of *education* are called for. The relation between these two objectives has in our day at times become a matter of contest—again of the "either-or" variety.

(4) LIBERAL EDUCATION AND VOCATIONAL TRAINING: FURTHER CONFRONTATIONS?

There is the radical professionalism of the empirical practitioners, who believe that robots well-trained in a given act, like circus dogs, are really all that a society needs, and obviously they have in view their own parochial image of society. And there are the humanists with a philosophical bend, who fancy that special expertises had better be left to emerge secondarily by some sort of fractional distillation from a primary background of broad general education—an education not for doing, but for critical thinking.

The most powerful argument posed by the latter group is that in order for a "democratic society", in the true sense of the word, to operate intelligently, every citizen should have his faculty for critical, i.e. rational, thinking and judging of issues developed to the maximum degree feasible for him; that he should be given some non-doctrinal idea of the nature of this world and of man in his various historic, geographic, functional, ideological, etc., groupings, so that he may gain some notion of what is basic and eternally inalterable in the *laws* of nature, and conversely, what has been, and will forever remain, subject to "adaptation", the evolutionary change of *styles* compatible with those laws. Purportedly, it is presumed that people thus "educated" would not only be better equipped for using judgement in their voting and political decisions, but also turn into more *judicious* nurses, draftsmen, craftsmen, salesmen, teachers, singers, and what not. But obviously, that would not be enough to make them also into *good* nurses, *good* craftsmen, and so forth.

So, what does it take to enable them to carry out their individual jobs and services not only judiciously, but truly competently and effectively. Not just something more, but something *else*, something that runs along a differently slanted axis of the educational body and that hence cannot be expected to emerge automatically from sheer general breadth of view and understanding. Rather than for breadth of perspective, it calls for depth of penetration, sharpness of focus, often practice by rote—in other words, a concentration on detail that is almost predicated on a deliberate confinement of view by blinders against diversion. So, if we were to compare the widening of vision and perspective, to which a general education aspires, to an expansion of the mind along a *horizontal* plane, the acquisition of competence in mastering the know-how in occupational specialties would have to be represented as a process in the *vertical* dimension, like the drilling of a shaft in depth. This simile might seem to contradict my caveat against polarizing issues in that it intimates that educational policy is truly faced with a decision between strictly dichotomized alternatives; "dichotomized" because the linearity of the time course of education—the tyranny of the clock—makes the two dimensions appear as if they were competitive in an algebraic sense. That such an "either-or" attitude is actually taken in educational circles, is amply documented by the endless debate about liberal education *versus* vocational training, as if the two were truly mutually exclusive.

In reality, of course, they are not. The case is not up for arbitration on points of expediency, but for resolution in terms of the *optimum harmonization* between the dual demands for the propitious self-development of the individual, on the one hand, and the exigencies of constructive service to

The Order of Life

society, on the other—that is, between "living" and "making a living". And when we turn from conventional glib designations to a critical analysis of the ingredients of the issue, we note at once that the introduction of linear time as a measure of educational limitations is, not wrong, but insufficient; what it lacks, are the considerations of *content* and of *rate*, inseparably connected with it. After all, education is a growth process, which must be characterized and measured not just by the time it takes to run its course, but by what it achieves in that course of time, and how efficiently; in other words, by the rate of gain of the growing body in proportion to time. And let me add that "gain" refers to an array of properties so numerous and diverse that no single feature could serve as a representative index of the "growth" of the whole system. In evaluating the growth of a living organism, the increments neither of its size, nor of its weight, nor of the number of its parts, nor of its functional proficiency, and so forth, could, taken by themselves, give us a fair measure of its consummate gain. If, furthermore, we bear in mind that food ingested does not contribute to growth unless and until it has been digested and assimilated, i.e., transformed into genuine tissue constituents—in other words, that the sheer addition of bulk cannot count as "growth"—we realize the spuriousness of our conventional quantitative criteria for educational advances, the measure of which could only be the total "gain" of that prominent feature of our organism: our mind.

Viewed under that aspect, bulk of "information" ceases to be a valid measure of "growth of knowledge" and even of "know-how", and the time required for the presentation and ingestion of a given quantum of informational data ceases to be a meaningful quantity in assessing the optimal rate at which an individual's knowledge and know-how can grow. When we transcribe this conclusion into the symbols of our dimensional simile, we realize that by assigning to the horizontal and vertical dimensions algebraic (=temporal) correspondence, we have been trapped into conventional ruts of thought. To get a correct mental model, we must visualize a *rectangle* whose sides are of the lengths of the horizontal line (for breadth) and of the vertical line (for depth), respectively;

the area which this rectangle encloses then represents "content of knowledge", and its expansion, "growth of knowledge". This mental diagram thus symbolizes the cohesiveness of the body of knowledge and the continuity and indissociability of its growth in both depth and breadth.

In terms of this image, we can now define more succinctly the task of education: it is *"to maximize the total content of the rectangle and to optimize the rate of its expansion."* This formulation has several corollaries. It illustrates the fact that the same result (rectangular content) can be achieved by an infinite number of conjugated pairs of values of the horizontal and vertical components, implying that the educational system need not freeze itself into a few rigid ratios between the two values, but should let the individual profit from their flexibility in finding the ratio best fitting his particular aptitudes, interests, aspirations, and circumstances. In practice, practicability sets certain limits to the realization of this ideal, but to replace the "tyranny of the clock" by the latitude conceded in our formula would significantly reverse the current educational trend, which aims at ever greater specialization, akin to the evolution of the insect state. In conclusion, I would decidedly answer the question at the head of this chapter in the sense that no categorical cleavage must be made between liberal and practical education, but that the only problem at issue in educational tactics is the change in the *proportion* in which they are to be apportioned in accordance with the varying rates of maturation of individuals and the progressive diversification of their occupational goals.

This is not to question the necessity of continuing the system of separate schools with different degrees of emphasis given to special intellectual or vocational interests. But it does negate the presumption that the assigning of greater weight, momentum and emphasis to special focal areas of application justifies the erection of artificial fences between them, ending in total mutual seclusion, alienation, and even competition. In short, as the needs of both society and its individual constituents require unremittingly the progressive tapering-in of intensive training for *know-how*, concern with the broad interlinking net of *general knowledge*, though necessarily diminishing in a reciprocal proportion, must never be allowed to fade out completely.

INDIVIDUALITY WITHIN THE LIMITS OF "CONTEXT"

In dealing, in the foregoing four sections, with the progressive decline in education of the breadth-to-depth ratio of the content of knowledge, we have by-passed the vital fact that this content itself is not a stationary quantity, but is—or rather, should be—kept in a state of perpetual growth from the cradle to the grave. The rate of this growth is again subject to the very wide range within which the different individual capacities, as well as the circumstances enabling them to come to full fruition and expression, vary. Yet, even though that rate would thus never submit to any universal formula, one can derive some general notions about it from the rules that apply to all growth of higher organisms. The one most pertinent in the present context stems from the fact that different parts and functions of the same body grow at different rates and also according to different time schedules. Proportionate growth is hardly ever observed. For example, if we compare the proportions of a baby with those of the adult, we find the head and brain of the baby disproportionately oversized relative to the rest of the body (note particularly the comparative shortness of the extremities), the disproportion being due to the precocity—a head start, as it were—of the former in embryonic development. At the same time, most of the latter parts grow through the fetal, infantile, and pubertal time course at a much faster rate, gradually overshooting the former.

The reason for presenting this example here as one for many is simply to illustrate the inequality of growth rates in development. It would be utterly misleading, for instance, if one were to go on now to extrapolate from the growth of the brain to the growth of the mind. The multiplication of our brain cells has essentially run its course shortly after birth, although the individual neuronal units (the individual nerve cells with their appendages and connections) can still keep enlarging in size and ramifications. But it is just about this period of subsidence of morphological brain growth, that the enormous functional enrichment of its innate repertory of perception and performance gets its spurt. My reference to "differential growth" in general, however, is crucial in the sense that it should dispel any expectation that the "content of the mind"—the rectangle of our model—ought to be allowed or made to grow proportionately in the dimensions of breadth and depth. It is my contention that while the faculties of animals gain by experience mostly in depth and refinement, man should be helped by education to favor the horizontal extension of his potential, that is, his breadth of vision with its corollaries of understanding, tolerance, and sane balance between reason and emotion. In short, I submit that even vocational training should aim, rather than at fashioning individuals in their impressionable years into human robots, at letting them grow into cultured human beings, who will have learned, despite their pre-occupation with specific tasks, to view those tasks in their critical bearing on the total cohesive fabric of "life on earth", with its precept for optimal combination of benefits both for the individual and, through him, for society and humanity.

In passing, let me side-track for a moment to a point of potential frustration from educational impotence. In the first place, the term "benefit" must in this connection be taken in its most exalted sense, including, besides mundane material receipts, the intangible spiritual rewards of moral satisfation. Now, an individual can usually assess rather clearly what he thinks to be of direct selfish benefit to himself. However, he can never decide with an equal feeling of certainty those benefits that might accrue to him as a beneficiary of membership in the hierarchy of collective groupings, real or contrived, of which mankind is composed—family, tribe, profession, nation, country, church, party, age group, or association from sheer bonds of friendship. Nor could he expect guidance from his sense of loyalty unless he were obsessed by monomaniac self-confinement to exclusive allegiance to only a single group relation in the listed complex. The task of a broadened education is precisely that of alleviating such propensity for letting one's decisions and actions be shaped by the straitjacket of exclusive, prejudicial, and inflexible attachment to a single group or cause—a narrowmindedness which

The Order of Life

the record of history has revealed to be a prime source of irrational discord. Therefore, people should be made aware early in life that obstinate single-trackedness of mind is apt to lead to injudicious decisions; whereas the soundness of judgement grows as one widens the range of aspects one takes into consideration and weighs as to their relevance.

However, the larger the range of considerations and points of view to be taken into account, the larger will also be the probability—and the dilemma—of intrapersonal contest between divergent beliefs and loyalties, breeding the discomfort of indecision. Imponderables have no objective weight, and yet they do weight each decision with value judgements, which can never be free from subjectivity. Family considerations may prevail over national concern or vice versa, national interests over professional loyalties or vice versa, ecumenical dreams over sectarian habits or vice versa, and so forth. The very fact that all such alternatives involve subjective value judgements defies the natural longing of people for hard-and-fast and safe rules that could act as guides for action regardless of circumstances; for it leaves it up to every individual's personal sense of responsibility to weigh and judge for himself each aspect of a given issue critically in the light of the *total relevant context* at hand and then determine in his own mind and conscience the *optimal balance* between benefit and cost ("cost-benefit ratio")—optimal in the sense of harmonizing self-interest with the diverse collective loyalties and responsibilities that I have hinted at above.

This amounts to a precept for approaching the resolution of problems and the settling of issues not by the simplistic practice of cutting down the assessment of potentially pertinent variables to the point where solutions begin to look easy, but on the contrary, by taking into account the widest range of possibly relevant aspects (the *"total relevant context"*)—a precept that aims at the gradual replacement of prejudice by responsible judgement. Unfortunately, the heavy burden of responsibility which this widening of scope places on the individual mind is generally resisted, even resented, and further retrenched by habits and conventions. Moreover, the narrowness of simplistic thinking is also grist for the mill of equally narrow political schemes, which court public favor by offering to an increasingly "security-hungry" populace the illusory prospect of "absolute security", not only economic, but also intellectual; the insecurity inherent in one's freedom for making personal judgements on the basis of critical thinking running of course counter to such political designs. For all those reasons, the maxim I have here advocated can be expected to be, on the whole, unpalatable and unpopular. This being so, and adopting the proposition as an essential ingredient for a sane and rational future of mankind, what stronger incentive could there be envisaged for education than to assume as one of its fundamental charges the obligation to reverse consistently the natural trend toward "simplisticism" in thinking by raising a generation with a frame of mind broad enough to act no longer single-mindedly in accordance with the restrictive codes of past "contexts" long since outdated, but with a realistic regard to the ever changing "ecology of contexts" that a perpetually changing world presents?

One of the didactic needs in that direction is a sharp turn from the ingrained thought pattern of linear "single cause-effect-chain" causality to the network causality of *"system dynamics"*, which would render the need for a "context concept" not only more compelling, but almost self-evident.

GROUP AND INDIVIDUAL: DOES "AVERAGE MAN" EXIST?

It would seem indicated, however, to bring at least one concrete example of why the introduction of the "context principle" deserves a prominent place in education—not so much for the guidance, as for the prevention of misguidance, of judgement. It refers to the fundamental disparity in valid statements about a *collective*, depending on whether the *collective as such* is at issue or whether extrapolations to its *members* are implied. A few pointers to the commonplace mischief perpetrated, innocently or cunningly, by ignoring or glossing over this distinction will show what I mean. The statistician's annual tallies of global agricultural produce over total human population yield some useful data about over-all trends, but any attempt to prorate them over the different individual members of the population would obviously be utter nonsense for it is precisely the purpose and resultant of any averaging procedure to level out or blur such individual distinctions. Now, in thermodynamics, stopping at statistical averages is adequately informative, for we are not interested in the peculiar, unique, non-recurrent excursions and gyrations of any particular separate molecule in a thermally agitated gas; what counts is only the behavior of the whole mass. Similarly, the exact tracing of the particular path along which any one person in an excited crowd is pummeled, would be rather uninteresting and uninformative. However, whether human beings go hungry because of a systematic local dearth of food, while there is a glut in other regions, though frightfully important to those affected, is lost in the averaging global tally, for "average man" just does not exist. Although it is quite useful and legitimate, in a given context, to operate with the fiction of "average man", in a different context, any resemblance between him and any real person would be a rare coincidence close to a miracle. Thus, there are contexts in which the mere numerical lumping of units, properties, aspects, etc., in "quasi-egalitarian" terms, must give way to regard for the *inequalities* among those items and the differential patterns of their distribution, so that the statistical results are no longer pertinent or applicable.

Now has this distinction ever been made compellingly and its profound lesson been driven home in educational practice? If it were, would we not witness fewer instances in which a perfectly just and valid statistical statement, covering a collective body, is illegitimately and illogically expanded to apply to the individual members of that body, as if an "average" could ever convey idiographic information? Does not the record of history and literature amply document the absurdly illogical and tragic consequences of this appraising and treating individuals not according to their personal qualifications but by the imaginary equal share allotted to every one of them from the calculated or merely presumed "average" properties of the collective to which they "belong"—a collective as small and circumscribed as to family strain, like the Capulets and Montagues, or as large and ill-defined as a given human tribe or race? And does not the final reconciliation of the Capulets and Montagues and the growing realization that meritorious diversity of human races is fully compatible with the recognition of the basic unity of all mankind, compatible as long as each proposition is considered in its proper context, carry a message? Does it not support my plea for turning a larger share of our educational planning from fire fighting—the dealing with overt crises—to fire prevention, that is, the raising of generations that will be alert to signs of inflammatory crisis-generating antagonisms rooted in disregard for disparities among the respective systemic contexts, and consequently will be intent and competent to defuse the incendiary charge by rigorous logic? I am presenting questions as food for thought. The answers are a matter for the future, not simply a matter of correcting missteps of the past.

This leads me over to a second example, no less far reaching and problematical, which will conclude this sampling of prospects for educational soulsearching. It has to do with the radical change in modern times of the rates—the kinetics—of change in the world around us, in which we are enmeshed. This reference to the extrinsic world is deliberate because I definitely do not want to insinuate that human nature—our inner world—is changing commensurately. The rate of external change, however, has reached a magnitude which has broken through the traditionally high level of imperceptiveness and lethargy of "average man", hence man's conservative feedback reactions to imminent or imagined disturbances of his dream of "status-quo security" are aroused. The resulting exacerbation of the already excessively polarized antithetical postures taken by advocates of an even more radical acceleration of rates of change, on the one hand, versus reactionary stoppage of change (e.g., "Zero Growth"), on the other—both equally unrealistic in concept and feasibility—need not be labored here; our whole social and political life bears its marks.

The Order of Life

TIME COURSE OF LIFE: ETERNAL CHANGE

Now, is this trend inexorable? Let us just take a glance at its sources, to which science has furnished major tributaries. In the early eighteenth century, it was still fashionable, even among many men of science, to cherish the belief in an everlasting static fixity or immutability of the content of the Universe. Even Heraclites' verdict of "Panta rhei" ("all is in flux") need not have shaken that belief for it could have been taken to imply simply incessant reshuffling of the same invariant content. But then the concept of evolution entered the scene, with all its scientific underpinnings in the history of siderial bodies, of organisms, of man and his ethnic, social, political and economic changes. Under the growing impact of this experience with the real course of the world, the idea of the inviolability of any "status quo" should have become increasingly suspect, had it not been for the comfort which "average man" is supposed to derive from the illusion of security in a stable world, in contrast to the sense of insecurity in a world of unpredictable change. The inertia instilled by this longing for absolute stability has permitted man to shut his eyes and mind to changes in and around him as long as the tempo of change was slow enough not to arouse attention and anxiety.

Now, in the recent past, the premises for this pleasant ostrich attitude have rapidly vanished. The great surge in the dissemination of information through the mushrooming visual and aural media of communication, intensified in its impact on the reader or listener by a certain bias in favor of "newsworthiness", i.e., impressiveness through shock value, has progressively invalidated the profession of ignorance as alibi for self-delusion and complacency. This, added to the dramatic acceleration of the actual pace of life and of the rate at which developments and innovations impinging on the individual occur, has all of a sudden shaken man out of his lethargy, has made him face change as a fact and learn to live with it—but "average man" just does not like it. And why? Because throughout history, including his own personal life, he has been inveigled into believing in the ideal of a status quo, the dream of a paradisical existence in a stagnant or stationary world undisturbed by change, instead of being made aware that "change", so-called, is a *primary* feature of the Universe as we have come to know it, and hence, must be accepted as a fact of life. The notion of constant stability is an abstraction that arises only in the context of processes in which the rate of change is slow enough to escape notice.

Man is unconscious of changes he cannot perceive. But he remains also undisturbed by certain changes of which he is aware, like the diurnal changes of light and darkness or the seasonal changes of weather or the rhythmic beating of his heart. He accepts them and gets to taking them for granted for one sole reason: his familiarity with the regularity of their rhythmic recurrence. The predictability of their reiterative return to what to him appears as a steady base line—a "status quo"—assures him of his cherished sense of security. Nonrecurrent changes, not previously experienced, bring insecurity, and hence, are feared. Yet, if he woke up to the fact that the reference base line, in the constancy and steadiness of which he trusts, is by no means itself unchanging, but is subject to continuous, though relatively slow, changes itself—e.g., the changes of climate on a geological time scale or the progressive changes of physiological rhythms with age, facts of which he does not become aware until they are brought to his consciousness, in the former case by instruction, in the latter case by signs of impaired health—he might get used to live in peace with the inexorable truth that progressive change is inherent in nature. He would no longer view unexpected occurrences as artificial and pathological disturbances of an imagined steady state of durability, but on the contrary, would recognize them as sheer outcroppings of that incessant stream of developmental processes of nature that flows uninterruptedly from the past into the future. He will come to acknowledge that unidirectional stream for what it is: just an aspect of the natural *time course* of the world, primary, basic, irrepressible, and irresistible; a steadily moving and metamorphosing carrier wave, of which the familiar rhythmic phenomena are but some transitory surface ripples. And what is more, he will acquire a sense of being part of that integral changing world—both as participant and tributary.

I firmly believe that, steeped in that way of thinking, man will no longer be alarmed, as he now often is, by signs of that stream of progress whenever he encounters them, but on the contrary, will become bewildered and disquieted by any moves that try to stop its flow. He will have learned that human energies had better be turned from the foolish and futile attempt at stemming a natural tide (of change) to rather channelling that force toward ends useful to man and mankind, or at least, preventing its dissipation or misdirection by ignorance, stupidity or eccentricity. To enhance this turn would merely take a change of viewpoint and of attitude, comparable to the change in the orientation of medicine from fighting disease to preserving health, or in the terms of a simile I have used before, the change of emphasis from firefighting to fireproofing.

Now, this task cannot be approached without a major wholehearted commitment to it of the educational process in all its stages. How to go about it, would be the next step in the row of open questions which I have raised in this book. To prepare answers would require far more extensive and penetrating deliberations, particularly by the experts in the tactics of the pedagogic art, than I would have the temerity to offer. Merely as a modest contribution to such deliberations, I wish to add the following remarks.

I sense that much of the habit of thinking in static terms is generated by the didactic practice of presenting data of information in finished and rather apodictic form. I realize, of course, the necessity of endowing a pupil with a solid grounding in the elements of knowledge without latitude for equivocation; surely, such items as the symbols and operational terms of mathematics, the rules of grammar, the measures of objects and processes of nature, certain dates as mnemonic anchor points for the chronology of history, and so forth, must first be hammered into his memory as elements of absolute immutability. Yet, let us bear in mind that such elements are just the piles driven in his mind on which to erect a consistent mental thought structure—auxiliary tools for a more precise description of the world than the ambiguous words of common language could provide; they are by no means in themselves descriptive of the course of the world, a world in perpetual flux of change. Thus, would it not help instill and strengthen in the pupils a genuine sense of the reality of this flux if most of the phenomena and events to which they are to be exposed were introduced to them not as isolated static items, but with some reference to their origins, developments, and even possible prospects, in a sort of evolutionary context? This need not be done in any elaborative way; often a mere hint would do. But the consummate effect of this procedure, if carried through systematically, would be to create and leave behind in the mind a natural feeling for the world as an entity in constant change, with an aversion to stagnation for its being unnatural.

It is obvious that if furthermore the same tactics were extended consistently to illuminating the connections and interrelations between *simultaneous* occurrences, this would enhance also the emergence in the maturing mind of the concept of *"relevant context"*, which I alluded to before, that is, the understanding of the integrative cohesiveness and inner connectedness of the world, with its systemic "network causality", as well as its "mysterious" coincidences. Such an understanding should greatly help the mind to accept more willingly the responsibility of judging and acting always with due regard for the variable "relevant context" of issues; which is, besides being a mandate for the present, also an indispensable prerequisite for a more rational planning by man of his future. And hopefully, the meaning of John Donne's reminder of more than three centuries ago that "man is no island entire of itselfe" will at last percolate through education to become a design for living instead of remaining just a literary curio.

EPILOGUE

I hope that the content of this postscript will likewise be judged in the relevant context for which it has been designed; that is, as a primitive attempt to raise to a higher level of perspective the sights for the urgently needed re-examination of the crucial objectives of education in a world that is in rapid motion without being quite sure whither and why. I have raised questions of such generality that they open themselves to a wide range of answers, ponderings, controversies, and further questions. My intent, quite open-ended, has only been to provoke thought and discussion, in the hope that it may, in the end, eventuate in concrete actions, not necessarily concordant, but at least convergent. To achieve this goal may take years; then, why delay a start? The task is clear: it is to change course from the fragmented, divergent, and dispersive dealings with the educational continuum piecemeal in tidbits, to a systematic concerted convergence upon the core problems of education, which center on my initial question of "what is the basic purpose of education?". To confound purpose with outcome by letting the purpose of education continue to be identified with simply the yield of the sum of the educational practices of the present, just will not do.

Evidently, the generality in which my questions are phrased is also justified by the multiformity of contexts (philosophical, social, political, religious, etc.) in which they will have to be answered, while yet converging upon a universal common goal from sources and along routes that not only are different but should continue to preserve and cultivate the constructive and creative features of their diversity. The answers from the Western world would presumably differ substantially from those of their Eastern counterparts; scientists, spending their conscious efforts in a domain of "demonstrable objectivity" will answer differently than will humanists or theologians, who will place greater stress on subjective value systems of morality and ethics; and so on. But there is so much in common to all human nature that in the end some common core of principles—a sort of common canon—of universal acceptability, and hence, powerful authority, would be distilled from all the diverse styles. Such a code could then serve as a guiding beacon for the orientation of education and civilized behavior all over the globe toward a sane humanism and human sanity.

The objection that time might be the limiting factor obstructing the implementation of practical answers to my questions is, to my mind, spurious. Contemporary curricula of education on all levels are so overstuffed with trivial data and inconsequential verbiage that the time necessary for the various innovations in the indicated directions could readily be freed by a radical switch from that diet of bulk and sweets, which has led to the current "information obesity", to a leaner and better balanced intellectual diet, rich in concentrates of items well selected for relevance and illustrative value for the building of a sound body of knowledge, rather than of human filing cabinets; in other words, by a judicious curtailment of redundancy.

Empiricists might blame or chide me for the slightly utopian slant of my view. This reminds me of the answer a colleague of mine, after delivering a convocation address, gave to the President of the University, who congratulated him on his speech but added that it seemed to him to have been somewhat "over the heads of the students". "Sir", replied my colleague, "I am always talking to the level where the heads of the students ought to be". To confine our vision to the level on which we already have answers at hand, does not seem to have been a very propitious attitude, even as it has enabled us to detect gaps and flaws in our educational past. The future calls for more.

To be effective, education will have to aim both at the prevention of further disintegration and, progressively, at a firmer consolidation of the unity of knowledge on which the sense of cohesion of human culture—singular—, coupled with full preservation of the creative diversity of human cultures—plural—, ultimately depends. Undoubtedly, the industrial age, with its technological profusion and with its growing pains of reconciling the aspiration to freedom for self-development of the human *individual* with the demands and restraints imposed on him by an evolving human society, is of itself not conducive to inaugurate, much less to nurture, such reintegrative moves. And yet, if mankind is to be kept from squandering man's unique dowry of reason by trading an

The Order of Life

incipient rational and humane trend of civilization for the stultifying prospect of a robotized and dehumanized world, science and the humanities, as now defined, will have to counteract their progressive self-isolation in self-grooving ruts by profuse anastomosing in networks of what nowadays is called "human concern". Such confluence calls for equilibration: more humanism in the humanities and less scientism in the sciences, till their value systems merge. I doubt whether we are faced as yet with "Two Cultures". But I can readily discern two growing *cults*.

The time has come when universities should reassert their time-honored role of leadership in adapting the lessons of the past to the exigencies of the future, Janus-faced. Universities are the major breeding and training grounds for teachers and future leaders of society. While having to spawn the mounting numbers of competent practitioners in the professions, arts and sciences, for whom modern society is clamoring, they must yet, at the same time, defend and fortify their station as trustees of knowledge, caretakers of the past and conquerors of the future. Accordingly, besides their obligation of training specialists for the professions and scholarly experts for the advancement of the diverse specialized branches of knowledge, they must, above all, be prime agents and exhibitors of the *unity, cohesiveness and inner consistency of all knowledge* to set a model for man of harmony instead of discord. In this sense, the blurring of such sharp academic distinctions as that between "science" and "humanities" becomes a mandatory task for education.

Fifteen years ago [18], I made a similar plea. Not much has changed in the interim. And so, I feel prompted to repeat that plea.

"Alarmed by signs that an abuse of science may lead to humans being treated as merely 'cases' for a gigantic statistical processing mill in which they are to be levelled to standards of the average, the common, and the mediocre, I make a plea to science to reacclaim diversity as source of progress (for uniformity means death), including the diversity of human minds in their responsible expressions. And then I make another plea to the nonscientific humanists not to regard themselves as prime custodians of civilization, shunning science as if it were inhuman. Let none of us lodge in the Master's Mansion, but let us all move down into the servants' quarters, so that we all may work together united for human progress in harmonious cooperation. The tasks are large, our forces limited. No group can do the job alone. So, let us all close ranks, the men of science with those in other walks of life, for humanism and against the dehumanization of our culture."

In the spirit of my call for "unity and continuity amidst diversity", which to nurture I have set forth as one of education's prime obligations, I wish to close with a little poem bearing on the subject, which I wrote some years ago. [19]

CONTINUUM

To life and time and space there is no end
But one continuous universal surge,
As seasons one into another blend,
Presses ahead in never ending urge
With unstaccatoed steadiness of trend;
"Now" but a flash where Past and Future merge.

All parcellations, marking phase or place,
Like merchandise in packages confined,
Are but devices of the human race
To trim immenseness to the grasp of mind—
Expedients to deal with time and space
By chopping entity to pieces uncombined.

Beginnings, ends—they are just artifact,
The knife marks of the trimming intellect;
Sign posts—no more—on an unending tract;
Bounds that we arbitrarily erect,
As if our own life, sensed as bounded act,
Could properly the Universe reflect.

Yet, raising sights from sample to the Whole,
I see the artificial boundaries fade.
I recognize my much more modest role
As but a flash in an eternal trade,
Moving from unknowns to an unknown goal,
Thread in a fabric not yet wholly made.

The Order of Life

To feel as part of this enormous stream,
Instead of just an isolated scoop,
As actor in a universal dream,
Will make your spirits soar, instead of droop.

You will not ask if life will ever stop
But you will sense that stoppage is a sham;
You will exult in living, drop by drop,
The verity: "I think; therefore I am."

Unbroken life lines passed down through the ages
Grant precedence to neither egg nor hen.
You note them just as serial sample stages
Of one continuum, including Man.

You will not ask if Winter's calm inaction
Means rest from Summer's consummated fruit
Or building strength for next year's satisfaction:
They are but signs of an eternal suit.

The Universe is one gigantic wave.
Its pulses are embodied in our lives.
We are its master neither, not its slave,
But part of it—that is, he is who strives.

The Present's arrow points to the Tomorrow,
Though what Tomorrow brings no one can say,
Except that, whether it be joy or sorrow,
TOMORROW surely will become TODAY.

May these thoughts strengthen the satisfaction of all human beings in being better today than they were yesterday, and inspire them to strive to become even better tomorrow than they are today, in the spirit of self-development, instead of trying to conform to the standards of fictitious "average man".

REFERENCES

References in the text to earlier works of the author that deal with the biological and epistemological foundations of the topics on "Wholes vs. Parts" (A), "Determinism" (B), and "System Causality" (C), and those that deal in detail with the facts of developmental (D) and cellular (E) biology on which the summary conclusions in the present book are based, refer to the following selected list of publications and, in a few cases, are further specified in the series of superscript numbers below.

(A) Weiss, Paul A. $1+1 \neq 2$. (When One plus One Does Not Equal Two) In: *The Neurosciences: A Study Program.* (Eds.: G. C. Quarton, T. Melnechuk and F. O. Schmitt). Rockefeller University Press, New York, pp. 801-821 (1967).

(B) _____ . The Living System: Determinism Stratified. In: *Beyond Reductionism—New Perspectives in the Life Sciences.* (Eds.: A. Koestler and J. R. Smithies). The Macmillan Company, New York, pp. 3-55 (1969).

(C) _____ . The Basic Concept of Hierarchic Systems. In: P. A. Weiss, *Hierarchically Organized Systems in Theory and Practice.* Hafner Publishing Co., New York, pp. 1-43 (1971).

(D) _____ . *Principles of Development: A Text in Experimental Embryology.* Henry Holt and Co., New York (1939); reprinted by Hafner Publishing Co., New York, LXXI and 601 pp., (1970).

(E) Weiss, Paul A. *Dynamics of Development: Experiments and Inferences.* Academic Press, New York, XIV and 624 pp., (1968).

Additional excerpts in the book have appeared before, in part, in "Life as Seen Through the Window of Life Science" (The Graduate Journal, The University of Texas, Supplement to Vol. VIII, pp. 88-156, 1970); "Depolarisation" (Studium Generale, Springer Verlag, Vol. 23, pp. 925-940, 1970); and "Impromptu on Education" (Studium Generale, Springer Verlag, Vol. 24, pp. 1377-1395, 1971).

[1] My original treatise on the subject (C) was published in German in BIOLOGIA GENERALIS, vol. 1, pp. 168-248 (1925). It was later translated into English and republished under the title of *"Animal Behavior as System Reaction"* in the YEARBOOK of the SOCIETY for GENERAL SYSTEMS RESEARCH, vol. 4, pp. 1-44 (1959). I concur with the Editor, who prefaced that republication after some thirty-five years by stating that "this paper is one of the earliest examples of systems-theoretical thinking in behavioral science from the biologist's point of view. . . .Naturally, Professor Weiss would not today subscribe to every contention and conclusion of the paper. . . .It is remarkable, however, that the basic tenets of the paper seem to have been so largely borne out by later develop-

ments." Although, unquestionably, my basic concepts of that time would nowadays be amenable to much refinement, particularly in the context of cybernetics, I am still amazed that the main framework has stood the test of time under the scrutiny of all my broader biological experiences since.

[2] Details on the subject may be found in (D), Part III.

[3] Summarized in (E), Chapter 2.

[4] Some further pertinent examples for the field-gene dualism are offered by the irreciprocity of hybrid crosses and by the so-called "hybrid merogones" (see (D), Part III, p. 183 ff.). The former refers to the fact that crosses between two distinct species often yield offspring with a decided prevalence of the maternal features attributable to the ooplasm, despite the equal contributions of genes from both parents. The merogone test is as follows. An unfertilized egg of species X is deprived by microsurgery of its own nucleus (genome X). It is then inseminated with sperm from a related, but quite distinct, species, Y. This composite egg (with ooplasm X and genome Y) is then left to develop into a young embryo, a patch of the cells of which is then transplanted into a normal embryo of the nuclear donor species, Y, thus placing the "hybrid" cells into an organism wholly of Y-character in both its genic and plasmatic constitution. The only genome then present in this animal, in the transplant as well as in the rest of the body, is the Y-type. Nevertheless, the tissue island developing from the graft produces and displays throughout life striking distinctive characteristics of species X, even though the latter had contributed merely enucleated ooplasm.

[5] (E) Chapters 8 and 11.

[6] (E) Chapter 4.

[7] (D) Part III, Chapters 3 and 4.

[8] (E) Chapter 6.

[9] (B) p. 12.

[10] (D) Part III, Chapter 1.

[11] E.g., (D), Figure 68, p. 328.

[12] E.g., (D), Figure 67, p. 326.

[13] (E) Chapter 2.

[14] (D) Part I, Chapter 2.

[15] (E) Chapter 1.

[16] This classification is based on the article cited in footnote 1.

[17] Examples in (D) Part III, Chapter 7.

[18] Weiss, Paul A. The Message of Science. Occasional Paper *1*, The Rockefeller University Press, (1959).

[19] Reprinted from Perspectives in Biology and Medicine, Vol. 12, No. 3, Spring 1969.